WORLD BANK STAFF WORKING PAPERS
Number 670

Patterns of Food Consumption and Nutrition in Indonesia

An Analysis of the National Socioeconomic Survey, 1978

Dov Chernichovsky
Oey Astra Meesook

The World Bank
Washington, D.C., U.S.A.

This is a working document published informally by the World Bank. To present the results of research with the least possible delay, the typescript has not been prepared in accordance with the procedures appropriate to formal printed texts, and the World Bank accepts no responsibility for errors. The publication is supplied at a token charge to defray part of the cost of manufacture and distribution.

The views and interpretations in this document are those of the author(s) and should not be attributed to the World Bank, to its affiliated organizations, or to any individual acting on their behalf. Any maps used have been prepared solely for the convenience of the readers; the denominations used and the boundaries shown do not imply, on the part of the World Bank and its affiliates, any judgment on the legal status of any territory or any endorsement or acceptance of such boundaries.

The full range of World Bank publications, both free and for sale, is described in the *Catalog of Publications*; the continuing research program is outlined in *Abstracts of Current Studies*. Both booklets are updated annually; the most recent edition of each is available without charge from the Publications Sales Unit, Department T, The World Bank, 1818 H Street, N.W., Washington, D.C. 20433, U.S.A., or from the European Office of the Bank, 66 avenue d'Iéna, 75116 Paris, France.

Dov Chernichovsky is with Ben Gurion University, Beer Sheva, Israel. Oey Astra Meesook is in the Country Policy Department of the World Bank.

Library of Congress Cataloging in Publication Data

```
Chernichovsky, Dov.
   Patterns of food consumption and nutrition in
Indonesia.

   (World Bank staff working papers ; no. 670)
   Bibliography: p.
   1. Nutrition surveys--Indonesia. 2. Food prices--
Indonesia. I. Oey Astra Meesook. II. Title.
III. Series.
TX360.I5C48  1984      363.8'2'09598      84-19555
ISBN 0-8213-0420-8
```

ABSTRACT

The purpose of this paper is to estimate the level of consumption of food and of nutrients for the Indonesian population; to identify population groups with nutrient deficiencies; to identify the major sources of different nutrients; and to estimate income and price elasticities of demand for both foods and nutrients.

The data source used is the National Socioeconomic Survey for 1978 (SUSENAS 1978) which included information on the quantities of, and expenditures on, about 120 food items consumed. Estimates of food and nutrient consumption of different population groups are given, with breakdowns by household expenditure class, region and area of residence. The proportion of the population with deficiencies in different nutrients is estimated by comparing each sample household's food consumption and the implied nutrient consumption against the household's own requirements, given the age composition of its members. The conclusion is that there are serious deficiencies in all nutrients in Indonesia and that the problem is more one of maldistribution than of an overall shortfall in the availability of foods. The problem is generally more serious in Java than in the Outer Islands, and affects the poorer households more severely than better-off households. In examining the sources of the various nutrients, the importance of rice as a contributor of most nutrients is striking.

The estimation of income and price elasticities of demand for food and nutrients is based on a household utility-maximization model from which the household's demand for food and hence nutrients is derived. For estimation purposes the double-logarithmic function is used, with quantities of food, calories and nutrients as dependent variables.

Variations in household incomes and in family size and composition are associated with marked variations in the quantities and patterns of food consumption; however, corresponding changes in the consumption of nutrients are much less dramatic. Thus the substitutions among food groups resulting from changes in income and family size and composition are fairly efficient in maintaining the nutritional intake of households.

The results concerning the effects of prices on food consumption patterns are rather tentative, since these prices reflect in part qualitative differences in food consumption which require further study. Nonetheless, it is already apparent that a great deal of substitution takes place as prices change.

The paper concludes that there is wide scope for nutrition policies based on changes in incomes and relative prices, as food and nutrition consumption respond rather dramatically to such changes. It is not clear at this stage how households respond to income and price changes in terms of the quality of foodstuffs they purchase, and what effect this has on nutrition. As incomes rise, the dependency on rice increases as well. However, any pricing policies designed to reduce this dependency must take into account the evidence suggesting that no single food can substitute for rice as a major source of most nutrients. The data also strongly suggest that inadequate diets are prevalent among the better-off and the better-educated as well. Hence, alleviating malnutrition in Indonesia is not just a matter of raising levels of income but also of nutrition education.

ABREGE

Cette étude a pour objet d'évaluer le niveau de consommation d'aliments et d'éléments nutritifs de la population indonésienne; d'identifier les groupes de population souffrant de carences nutritionnelles; d'identifier les principales sources des divers éléments nutritifs; et d'estimer les élasticités – revenu et prix de la demande des aliments et des éléments nutritifs.

Les données proviennent de l'Enquête socioéconomique nationale de 1978 (SUSENAS 1978) qui comportait des renseignements sur la quantité et le prix de quelque 120 produits alimentaires consommés. La présente étude donne des estimations des aliments et éléments nutritifs consommés par les divers groupes de population, ventilées par classe de dépense des ménages, région et lieu de résidence. On a calculé le pourcentage de la population souffrant de carences nutritionnelles de diverses sortes en comparant la consommation d'aliments, et donc indirectement celle d'éléments nutritifs, de chaque ménage échantillon et les besoins de ce ménage, compte tenu de sa composition par âge. Cette comparaison permet de conclure qu'en Indonésie il existe des carences très importantes de tous les éléments nutritifs, mais qu'elles sont dues plutôt à une mauvaise distribution qu'à une pénurie généralisée de produits alimentaires. D'une manière générale, le problème est plus grave à Java que dans les autres îles et les ménages pauvres sont plus touchés que les ménages aisés. Lorsque l'on étudie la source des divers éléments nutritifs, l'importance du riz est frappante.

Pour évaluer les élasticités - revenu et prix de la demande d'aliments et d'éléments nutritifs, on a utilisé un modèle de maximisation d'utilité des ménages qui a permis de déterminer la demande d'aliments et donc d'éléments nutritifs des ménages. Aux fins du calcul, on a utilisé une fonction à double logarithme où les quantités d'aliments, de calories et d'éléments nutritifs apparaissent comme variables dépendantes.

Les variations des revenus des ménages ainsi que de la taille et de la composition de la famille vont de pair avec des variations notables des quantités d'aliments et des habitudes de consommation; mais les modifications correspondantes de la consommation d'éléments nutritifs sont beaucoup moins marquées. Les substitutions entre les groupes d'aliments permettent donc aux ménages de maintenir de manière relativement efficace leur ration nutritive, quelque soit le revenu et la taille et la composition de la famille.

Les conclusions portant sur les effets des prix sur les habitudes de consommation sont relativement ambiguës puisque ces prix sont en partie le reflet de différences qualitatives, qui doivent être étudiées plus avant. Il apparait néanmoins que les fluctuations des prix se traduisent par des substitutions importantes.

La conclusion de cette étude est que l'on peut envisager un éventail assez large de politiques nutritionnelles fondées sur des modifications des revenus et des prix relatifs, puisque la consommation d'aliments et d'éléments nutritifs répond de manière spectaculaire à des modifications de ce genre. On ne sait pas encore exactement les effets que les modifications des revenus et des prix ont sur la qualité des aliments achetés par les ménages, et pour conséquent, sur la nutrition. A mesure que les revenus augmentent, les ménages mangent de plus en plus de riz. Cependant toute politique des prix ayant pour objet de réduire cette dépendance doit tenir compte du fait qu'aucun produit ne peut remplacer à lui seul le riz comme source d'éléments nutritifs. Les données recueillies semblent aussi suggérer que même les régimes alimentaires des ménages aisés et instruits sont mal équilibrés. Pour réduire la malnutrition en Indonésie, il ne faudra pas seulement relever le niveau des revenus mais aussi celui de l'éducation alimentaire.

EXTRACTO

La finalidad de este documento es estimar el nivel de consumo de
alimentos y nutrientes de la población de Indonesia; identificar los
grupos de población afectados por deficiencias de nutrientes; señalar las
fuentes principales de los diferentes nutrientes, y estimar la elasticidad
de la demanda de alimentos y de nutrientes tanto en función del ingreso
como de los precios.

La fuente de datos utilizada es la Encuesta Socioeconómica
Nacional de 1978 (SUSENAS 1978), que contiene información sobre las
cantidades consumidas de unos 120 artículos alimentarios y sobre los
gastos correspondientes. Se presentan estimaciones del consumo de
alimentos y nutrientes de diferentes grupos de población, desglosadas por
categoría de gastos familiares, por región y por zona de residencia. La
proporción de la población afectada por deficiencias de diferentes
nutrientes se ha estimado comparando el consumo de alimentos y el consumo
implícito de nutrientes de cada familia de la muestra con las necesidades
reales de la familia, teniendo en cuenta la composición de la misma por
edades. La conclusión a que se llega es que en Indonesia hay graves
deficiencias de todos los nutrientes y que este problema radica más en la
mala distribución que en la insuficiente disponibilidad de alimentos en
general. El problema suele ser más serio en Java que en las islas
exteriores y afecta en mayor grado a las familias más pobres que a las más
pudientes. Al examinar los orígenes de los diversos nutrientes es muy
notable el predominio del arroz como elemento participante de muchos de
ellos.

La estimación de la demanda de alimentos y nutrientes en función
del ingreso y de los precios se basa en un modelo de maximización de la
utilidad familiar del que se deriva la demanda de alimentos y, por tanto,
de nutrientes de la familia. Para elaborar la estimación se ha usado la
función logarítmica doble, tomando como variables dependientes las
cantidades de alimentos, calorías y nutrientes.

Las variaciones en los ingresos familiares y en el tamaño y
composición de la familia dan lugar a marcadas variaciones de la cantidad
y modalidad de consumo de alimentos, pero las modificaciones
correspondientes del consumo de nutrientes son mucho menos marcadas. Así,
las sustituciones entre grupos de alimentos resultantes de las variaciones
del ingreso y del tamaño y composición de la familia contribuyen en forma
bastante eficiente a mantener el nivel de ingestión de nutrientes de las
familias.

Los resultados concernientes a los efectos de los precios en la
modalidad de consumo de alimentos son bastante tentativos, dado que estos
precios reflejan en parte diferencias cualitativas del consumo que exigen
un estudio más a fondo. No obstante, ya es evidente que a medida que los
precios varían tiene lugar una considerable sustitución de alimentos.

En el documento se llega a la conclusión de que hay amplio
margen para adoptar políticas nutricionales basadas en las variaciones de
los ingresos y de los precios relativos, dado que el consumo de alimentos
y la nutrición reaccionan en grado bastante espectacular a esas
variaciones. En la actualidad no se sabe exactamente cómo reaccionan las
familias ante las variaciones de los ingresos y los precios en cuanto a la
calidad de los productos que compran, ni qué efecto tiene esto en la

nutrición. Cuando los ingresos aumentan, también aumenta la dependencia del arroz. Sin embargo, cualquier política de precios tendiente a reducir esa dependencia debe tener en cuenta los datos que indican que ningún alimento, por sí solo, puede sustituir al arroz como una de las principales fuentes de la mayoría de los nutrientes. Además, los datos señalan claramente que las dietas inadecuadas también prevalecen entre los grupos de mayores recursos y nivel de instrucción. Por lo tanto, para aliviar el problema de la malnutrición en Indonesia no sólo se requiere elevar el nivel de ingresos sino también mejorar la educación en la esfera de la nutrición.

Table of Contents

ACKNOWLEDGMENTS

This paper was prepared under the World Bank's Research Project 672-19, "Poverty, Fertility and Human Resources in Indonesia". The authors would like to express their appreciation to the Central Bureau of Statistics of Indonesia for making available the SUSENAS 1978 data tapes to the project.

Patterns of Food Consumption and Nutrition in Indonesia

Dov Chernichovsky
Oey Astra Meesook

I.   INTRODUCTION

Indonesia is generally thought to have a nutrition problem among its population which, until recently, has been associated with shortages of rice, Indonesia's basic staple. The Food and Agriculture Organization (FAO) of the United Nations has estimated that the nutritional status of a significant portion of the population, especially of children, is below standard. [1] Along with basic calorie and protein deficiencies, micronutrient deficiencies of vitamin A, iodine and iron are also believed to be prevalent.

Because of lack of suitable data, no direct link has yet been established between mortality, morbidity, and nutritional status, on the one hand, and food consumption, nutrition, and income, on the other. [2] The problem of malnutrition in Indonesia can be traced to poverty and lack of education on

---

[1]   Only scanty information is available on this matter. See S. Tabor, "SUSENAS V - Preliminary Evaluation of Consumption Trends and Nutritional Status", mimeo, USDA, Washington D.C., 1979; T.J. Ho, "Economic Status and Nutrition in East Java", World Bank, Washington D.C., mimeo.

[2] See T.J. Ho, op. cit.

nutrition. It is compounded by a high degree of dependence on rice, of which Indonesia is a major consumer and thus can influence the world price.

The above state of affairs has led to the formulation of three major policy objectives: -

(i)   An overall increase in both the production and consumption of food.

(ii)   Diversification of the diet to avoid excessive reliance on any one staple, particularly rice.

(iii)   Stabilization of food consumption of vulnerable groups: the poor, pregnant and lactating women, and children.

As far as the first policy objective goes, Indonesia has made substantial gains in food production. The production of rice has increased at about 4.0% per annum over the last decade, while population grew at about 2.4% per annum, implying an increase in the average per capita consumption of domestically-produced rice. At the same time, however, the reliance on rice as the major staple increased; it represented 72% of total staple consumption in 1980, compared with 68% during the 1970-75 period. This was probably related to changes in incomes and the pricing policies of the 1970's in which the price of rice was kept low relative to the prices of other major staples such as cassava, wheat and corn.

It has been pointed out that total food availability in Indonesia is now more than sufficient to meet average nutrition requirements of the population. [1] How this increase in total food availability translates into improvements in food and nutrient intake depends on its distribution among the population.

The effect of changes in incomes and relative prices as a result of improvement in food production on the consumption levels and patterns of different population groups is a critical issue in the design of food policies and programs. The issue relates to the households' decision-making process insofar as it affects nutritional intake. The response of households to changing opportunities may well be inconsistent with the basic policy objectives and with the improvement in the level of welfare of the population as perceived by the policy makers.

The objective of this paper is to study the relationship between food and nutrient consumption and household characteristics in Indonesia. It is proposed to estimate the level of consumption of food and nutrients for the population; to identify population groups with nutrient deficiencies; to identify the major sources of different nutrients; and to estimate income and price elasticities of demand for both foods and nutrients.

The data source for the study is the National Socio-Economic Survey for 1978 (SUSENAS 1978). The next section gives details of the data set and presents estimates of food and nutrient consumption of different population groups, with breakdowns by household expenditure class, region, area of residence, and whether the household is deficient in calories, protein and vitamin A. The estimates are available for different seasons of the year and these are shown as well. In addition, this descriptive material is used to identify

---

[1] World Bank, Report No. 3795-IND, <u>Indonesia: Financial Resources and Human Development in the Eighties</u>, May 1982. (This is an internal document with restricted circulation.)

the groups with consumption of nutrients below the recommended levels and hence are most vulnerable to nutrition-related problems.

Section III gives some conceptual and analytical considerations which are followed in Section IV by estimates of income and price elasticities of demand for foods and nutrients, information which would be critical for the design of income and price policies to improve food consumption patterns and the level and pattern of nutrition in Indonesia.

II. THE DATA AND THE SETTING

2.1 The Data

The data used in this study are from the data tapes of the National Socio-Economic Survey (SUSENAS) conducted by the Central Bureau of Statistics in 1978. The survey was carried out in four rounds each covering a three-month period; these were centered around February, May, August, and November. The sample size was over 6,300 households in each round, with a total for all four rounds of over 25,000 households. Each observation of each of the subsamples is treated separately in this study. [1/] This approach combines the panel characteristics of the data that are important for variations in prices, with cross-sectional properties. The basic characteristics of the sample are presented in Table 1.

The regional and urban/rural distributions of the sample, shown in columns 1 and 2, are almost identical to those of the 1971 Population Census and the 1976 Intercensal Population Survey. Nearly two-thirds of the

---

[1/] Because of problems with the data tape for the August round, only the remaining three rounds were used in the analysis. Accordingly, figures showing average values are averages for the three rounds only.

Table 1: BASIC SAMPLE CHARACTERISTICS, SUSENAS 1978 DATA, INDONESIA, 1978

Region	Population (%)	Rural Population (%)	Average Household Size (persons)	% Age 10 & above (%)	Household monthly expenditure [1] (Rp)
Java	65.76	83.67	4.6	71.8	23,574
DKI Jakarta	3.55	0.00	5.7	71.9	82,197
West Java	20.38	89.27	4.4	70.5	24,795
Central Java	17.98	90.00	4.8	70.8	17,183
DI Yogyakarta	2.09	84.72	4.7	76.6	21,840
East Java	21.76	86.84	4.5	73.3	18,585
Outer Islands	34.24	82.48	5.2	67.6	31,437
Sumatra	17.22	81.79	5.3	67.3	33,309
Bali & Nusatenggara	5.38	92.65	5.0	68.0	22,719
Kalimantan	4.63	75.30	5.0	68.0	34,120
Sulawesi	6.72	81.63	5.3	67.9	31,582
Maluku & Irian Jaya	0.29	58.12	5.7	68.4	40,799
Indonesia	100.00	83.39	4.8	70.8	26,233

**   Exchange rates applicable for the period are

$ 1 = Rp 415 before November 15, 1978;
$ 1 = Rp 625 after November 15, 1978.

Source: SUSENAS 1978 data tapes, Biro Pusat Statistik, Jakarta.

[1] Based on the second round of the survey conducted in May 1978.

population are concentrated on Java. Outside of the capital city of Jakarta,
most of the remainder of Java is classified as rural. Rural areas contain a
slightly smaller proportion of the total population in the Outer Islands. The
SUSENAS sample thus provides a representative geographical distribution of the
population of Indonesia. [1]

Average household size, shown in column 3, is 4.6 persons overall.
Javanese families tend to be smaller than families in the Outer Islands. This
phenomenon, associated with Java's relatively low fertility rate, is consistent with the figures in column 4, which show a higher proportion of persons
age 10 and above in Java than in the Outer Islands. [2]

Household monthly consumption expenditure will be the basic welfare
indicator used in this study. Column 5 presents this indicator which varies
substantially from around Rp 17,000 in Central Java to over Rp 30,000 in
Sumatra, Kalimantan and Sulawesi. Jakarta shows a very high average expenditure. Likewise Maluku and Irian Jaya where the urban population is overre-
presented in the sample. Average monthly expenditures are lower in Java than
in the Outer Islands, and they also show more variation. This is further
shown in Table 2 in which expenditure ranges are given which divide households
into the lower 40%, middle 30% and upper 30%. The classification is done

---

[1]   See World Bank, _Employment and Income Distribution in Indonesia_, World
Bank Country Study, Washington, D.C., July 1981, Table 1.1, p.4, and Appendix
Table 1.

[2]   For specific regional demographic and economic characteristics, see Dov
Chernichovsky and Oey Astra Meesook, _Regional Aspects of Family Planning and
Fertility Behavior in Indonesia_, World Bank Staff Working Paper No. 462, May
1981.

Table 2: CLASSIFICATION OF HOUSEHOLDS BY TOTAL MONTHLY CONSUMPTION EXPENDITURES, INDONESIA, 1978

Expenditure Class	Range of Household Monthly Expenditures (Rp)	
	Java	Outer Islands
Lower 40%	up to 12,999	up to 22,999
Middle 30%	13,000-23,999	23,000-35,999
Upper 30%	24,000 or more	36,000 or more

Notes: The bottom category actually contains 39.2% of the population in Java and 40.2% in the Outer Islands; the middle category contains 30.8% and 29.5% respectively.

Source: SUSENAS 1978 data tapes, Biro Pusat Statistik, Jakarta.

separately for Java and the Outer Islands; it can be seen that the bottom 40% in the Outer Islands is more "affluent" than that in Java.

2.2 The Setting

The data, of which we use the months of February, May and November, provide information on the quantities of, and expenditures on, approximately 120 food items. To simplify the discussion, we have grouped similar food items into thirteen major groups as follows: rice, corn, wheat, cassava, potatoes, fish, meat and poultry, eggs, dairy products, vegetables, legumes, fruits, and other. An attempt was made to keep the items in each group as homogeneous as possible except that the last category is basically a residual one, containing those items not found in any other group. 1/ However, the nutritional values of foods were calculated on the basis of individual food items. 2/ An adjustment was made for food and nutrient consumption from "food and beverages outside the home" based on the household's expenditures on these in relation to the expenditures on food consumed at home.

Price information was not explicitly available in the data. Average implicit prices were obtained by dividing the expenditure on any particular item by its quantity (whenever positive values and quantities were reported.) This issue will be discussed further in the next section.

1/ For a detailed list of items in each food group, see Annex 1.

2/ The conversion was based on two highly consistent sources: United States Department of Agriculture, Nutritive Value of Foods, Home Garden Bulletin No. 72, Washington D.C., 1971 and FAO and USHEW, Food Composition Tables for Use in South East Asia, Washington D.C., 1972.

2.2.1 Food Consumption Patterns

The shares of expenditures on major consumption categories in total household expenditures are shown in Table 3. On average 68% of the household's total expenditures are spent on food. Across regions, Jakarta has the lowest food share and Kalimantan the highest. The rural population spends relatively more on food than the urban population, an indication of the relative affluence of the urban dwellers. Seasonality does not have much effect on the food share.

As expected, the food share falls with rising incomes. While the poorest 40% of households allocate 73% of their total expenditures to food, the richest 30% allocate 59%. It is possible that the latter figure is an overestimate if the upper income groups underreport nonfood expenditures.

Table 3 shows the variations in the allocation of the household budget to different expenditure categories with rising incomes. The budget shares of all nonfood categories generally, and in particular "goods and services", "durable goods", schooling and health-related expenditures which are "luxury items", increase with rising total expenditures.

Table 4 gives the proportions of the household food budget allocated to each food group. For the total population, rice accounts for one-third of total food expenditures, other staples for another 7%; fish, meat and poultry for 9%; eggs and dairy products for 2%; vegetables, legumes, and fruits for 13%; and other items which include oil, butter, sugar, bread and drinks for the remaining 35%.

Table 3: PROPORTIONS OF TOTAL HOUSEHOLD EXPENDITURES ALLOCATED TO MAJOR CONSUMPTION CATEGORIES BY REGION, LOCATION, SEASON, AND EXPENDITURE CLASS, INDONESIA, 1978

(%)

	Food	Alcohol & Tobacco	Housing & Fuel	Clothing & Footwear	Goods & Services	Schooling	Health	Durable Goods	Taxes & Insurance	Parties & Ceremonies
Indonesia	68.00	4.92	12.32	5.10	4.27	1.10	1.09	1.86	1.20	2.59
Region										
DKI Jakarta	51.19	5.30	22.11	4.64	12.71	2.88	1.84	1.64	1.47	1.11
West Java	71.04	4.05	12.89	4.68	3.71	0.90	1.01	1.73	0.84	1.31
Central Java	65.92	4.48	13.30	4.92	4.51	1.18	1.24	1.75	1.88	3.63
DI Yogyakarta	59.23	4.25	12.73	6.75	8.62	2.89	2.13	2.78	1.35	4.51
East Java	64.50	4.96	13.64	4.89	4.36	1.14	1.32	2.12	1.10	4.99
Sumatra	70.14	6.26	9.60	5.75	4.01	1.13	0.89	2.13	1.04	1.32
Bali & Nusatenggara	73.41	4.33	9.78	4.75	2.83	0.81	0.77	1.31	1.26	2.59
Kalimantan	72.84	6.36	10.01	4.83	2.90	0.64	0.64	1.59	0.52	1.20
Sulawesi	72.56	4.42	9.30	5.95	2.74	0.64	0.65	1.85	1.97	1.46
Maluku & Irian Jaya	67.38	4.65	12.14	6.69	6.13	1.86	0.54	1.65	0.75	0.85
Location										
Urban	59.72	5.36	17.43	4.97	8.52	2.41	1.58	1.81	1.14	1.29
Rural	69.63	4.83	11.31	5.13	3.43	0.84	0.99	1.87	1.21	2.85
Season										
February	68.41	4.97	13.11	4.32	4.55	1.40	1.13	1.72	0.99	2.18
May	68.49	5.23	12.27	4.81	4.46	1.04	1.16	2.00	1.24	1.75
November	67.04	4.55	11.59	6.17	3.81	0.87	0.98	1.87	1.37	3.84
Expenditure Class										
Lower 40%	72.70	4.46	13.24	3.89	2.21	0.47	0.65	0.53	1.04	1.91
Middle 30%	69.91	5.19	11.08	5.11	3.79	0.95	0.99	1.16	1.15	2.62
Upper 30%	59.07	5.15	12.33	6.63	7.51	2.09	1.75	4.38	1.46	3.42

Source: SUSENAS 1978 data tapes, Biro Pusat Statistik, Jakarta.

Table 4: PROPORTIONS OF THE FOOD BUDGET ALLOCATED TO DIFFERENT FOOD GROUPS, INDONESIA, 1978

(%)

	Rice	Corn	Wheat	Cassava	Potatoes	Fish	Meat & Poultry	Eggs	Dairy Products	Vegetables	Legumes	Fruits	Other
Indonesia	33.90	3.65	0.61	1.88	0.74	6.56	2.48	1.07	0.72	7.30	3.16	2.53	35.40
Region													
DKI Jakarta	21.52	0.09	0.13	0.28	0.46	5.30	4.06	2.59	2.73	6.90	4.15	3.74	48.05
West Java	44.24	0.72	0.19	1.09	0.57	6.80	2.15	0.91	0.66	5.55	2.76	3.09	31.27
Central Java	33.73	7.40	0.48	2.44	0.51	2.38	1.19	0.64	0.47	7.98	3.80	1.63	37.36
DI Yogyakarta	25.93	2.11	0.29	2.90	0.20	0.47	0.83	0.97	0.69	9.31	3.81	1.50	50.99
East Java	27.79	7.34	1.28	3.32	0.43	4.68	2.16	1.01	0.41	7.95	4.96	1.67	37.01
Sumatra	35.00	0.15	0.32	1.10	0.99	9.94	3.36	1.35	0.96	8.50	1.79	2.81	33.72
Bali & Nusatenggara	34.39	5.46	0.96	1.56	1.82	5.42	5.40	0.99	0.28	7.03	1.93	2.82	31.93
Kalimantan	30.11	0.23	0.51	1.12	0.52	12.99	3.06	1.39	1.03	6.90	1.83	3.49	36.82
Sulawesi	31.87	2.95	0.94	1.34	1.67	13.18	2.70	1.19	0.93	5.67	1.54	4.02	32.00
Maluku & Irian Jaya	17.29	0.13	0.71	4.98	4.01	15.40	1.65	1.08	2.50	9.11	2.20	3.74	37.19
Location													
Urban	27.89	0.40	0.25	0.50	0.53	6.90	3.91	2.03	2.03	7.20	3.85	3.39	41.12
Rural	35.11	4.31	0.68	2.16	0.78	6.50	2.19	0.87	0.45	7.32	3.02	2.36	34.25
Season													
February	32.97	5.45	0.99	2.04	0.72	6.34	2.13	0.97	0.73	7.11	3.08	2.80	34.65
May	33.77	3.02	0.43	1.70	0.66	6.73	2.13	1.12	0.66	7.67	3.23	2.48	36.40
November	35.02	2.43	0.40	1.91	0.84	6.62	3.23	1.10	0.76	7.10	3.17	2.29	35.13
Expenditure Class													
Lower 40%	36.29	6.41	0.82	2.80	0.77	5.88	0.86	0.59	0.14	8.04	2.75	1.67	32.99
Middle 30%	36.86	2.68	0.48	1.65	0.69	6.75	2.04	0.96	0.49	7.05	3.14	2.42	34.78
Upper 30%	28.01	1.15	0.47	0.96	0.74	7.24	4.94	1.77	1.66	6.62	3.68	3.71	39.04
Nutritional Status													
Calories: Deficient	34.73	5.50	0.68	1.73	0.60	5.64	1.71	0.37	0.53	7.40	3.31	3.28	35.39
Not Deficient	32.91	1.45	0.52	2.07	0.90	7.66	3.40	1.29	0.93	7.19	2.97	1.90	35.42
Protein: Deficient	35.43	6.52	0.67	2.78	0.65	4.41	1.06	0.66	0.28	7.58	2.75	3.14	35.62
Not Deficient	32.89	1.76	0.57	1.29	0.80	7.98	3.42	1.33	1.00	7.12	3.44	1.60	35.26
Vitamin A: Deficient	38.75	3.54	0.55	1.80	0.35	6.32	1.86	0.79	0.45	6.02	3.01	3.16	34.78
Not Deficient	29.78	3.75	0.66	1.95	1.07	6.77	3.01	1.30	0.94	8.39	3.28	1.78	35.94

Source: SUSENAS 1978 data tapes, Biro Pusat Statistik, Jakarta.

The urban population spends proportionately less on rice, corn, and cassava (the main staples) than does the rural population, and more on meat and poultry, eggs, milk, legumes, fruits, and "other" foods. Similarly, the lower 40% allocates more of its budget to the staple foods than do the other expenditure groups, while the upper 30% spends relatively more on fish, meat and poultry, eggs, milk, and "other" foods. Within staples, the rural population spends a smaller proportion on rice and larger proportions on corn and cassava. The share of staples in the food budget declines while the share of rice within the staple group increases with rising incomes. This has clear implications for the aggregate demand of these major crops with rising incomes over time, and is consistent with the observed rising dependence on rice in Indonesia.

Some minor variations in the pattern of food consumption can be seen by season. For example, less is spent on rice and more on corn and wheat in February compared with November.

The major differences in the pattern of food expenditures are across regions. For example, the share of rice in the total food budget ranges from 17% in Maluku and Irian Jaya to 44% in West Java, while the share of cassava ranges from close to zero in Jakarta to 5% in Maluku and Irian Jaya. In all major food groups there are substantial variations across regions.

A comparison between the population groups defined according to whether or not they are deficient in different nutrients shows that those groups which are deficient in calories and protein spend higher proportions of their food budgets on staples, such as rice and corn, and lower proportions on fish, meat and poultry, eggs and dairy products.

Table 5 shows the proportion of the population reporting <u>some</u> expenditure on each food group. Rice, vegetables, and "other" food items are consumed by the vast majority of the population. A substantial proportion of the population also reports consumption of fish, legumes, and fruits. Wheat and dairy products are reportedly consumed by only 8% and 14% of the population respectively.

For most food groups, consumption is reported by a larger proportion of the urban population than of the rural population. Thus the urban population has on average a more varied diet, one which includes food items from a larger number of food groups, compared with the rural population. The exceptions to this are corn and cassava. Corn is consumed by 26% of the rural population compared with 9% of the urban population; and 47% of the rural population consume cassava compared with 33% of those in urban areas.

As expected, additional income is associated with a greater likelihood of consuming most foods. Corn, however, appears to be an inferior good; the higher the expenditure group, the lower the proportion of the population that consumes it. A larger proportion of the population who are deficient in calories, protein and vitamin A consume corn than those who do not have these deficiencies.

One should also note the exceptionally low proportion of households in Central Java and Yogyakarta reporting consumption of fish, meat and poultry, and dairy products.

For households reporting consumption, Table 6 shows per capita daily consumption in grams (except for eggs which are given in units) of the various food groups by region. Consumption of foods seems to be more similar among the population of the Outer Islands than among the Javanese. For example, per

Table 5: PROPORTIONS OF HOUSEHOLDS REPORTING CONSUMPTION OF DIFFERENT FOOD GROUPS, INDONESIA, 1978
(%)

	Rice	Corn	Wheat	Cassava	Potatoes	Fish	Meat & Poultry	Eggs	Dairy Products	Vegetables	Legumes	Fruits	Other
Indonesia	94.37	23.61	8.30	44.82	23.91	81.80	21.77	31.10	14.01	97.44	65.65	54.40	99.62
Region													
DKI Jakarta	99.52	6.68	14.17	29.09	37.27	90.61	52.56	66.73	52.31	97.10	94.37	76.12	99.21
West Java	99.83	16.48	3.47	52.03	25.48	93.34	17.64	24.27	12.30	95.45	68.02	61.45	99.33
Central Java	85.47	30.87	2.91	43.40	15.04	63.80	12.11	17.87	8.11	98.88	70.58	39.90	99.55
DI Yogyakarta	96.03	16.15	1.46	52.10	12.62	22.45	11.95	28.28	13.31	97.75	87.69	45.33	99.94
East Java	88.74	39.62	9.13	48.30	16.87	74.17	20.21	30.66	6.16	99.31	83.01	38.85	99.84
Sumatra	99.80	4.49	10.46	39.84	35.92	95.95	26.54	40.93	21.96	98.72	50.79	63.05	99.74
Bali & Nusatenggara	95.18	37.06	7.81	33.55	27.76	70.97	37.00	30.98	5.53	90.92	44.83	59.44	99.93
Kalimantan	99.96	7.22	16.98	53.05	24.31	96.27	28.66	41.99	22.05	98.38	44.55	70.17	99.88
Sulawesi	98.09	35.57	22.06	36.59	27.05	95.23	23.06	38.33	19.98	94.87	37.11	76.19	99.37
Maluku & Irian Jaya	97.18	4.68	21.69	68.67	46.11	97.99	17.70	26.18	45.47	98.96	47.48	60.93	99.57
Location													
Urban	99.59	9.63	12.44	33.18	32.50	83.52	42.74	54.99	38.25	97.21	81.48	68.41	99.24
Rural	93.32	26.43	7.47	47.17	22.18	81.45	17.55	26.29	9.13	97.49	62.46	51.58	99.70
Season													
February	91.22	38.50	10.23	40.81	21.95	80.87	19.13	28.27	13.99	97.48	63.01	52.73	99.65
May	95.03	17.07	6.72	41.49	22.24	82.11	19.65	31.89	13.49	97.38	65.99	55.33	99.49
November	96.99	14.89	7.95	52.61	27.75	82.45	26.83	33.25	14.59	97.47	68.08	55.17	99.72
Expenditure Class													
Lower 40%	88.18	27.99	4.30	41.83	14.64	73.45	6.35	14.41	1.92	95.80	53.23	36.67	99.33
Middle 30%	97.48	22.49	6.69	46.72	22.06	84.85	16.90	29.11	8.04	97.98	66.75	54.25	99.88
Upper 30%	99.09	19.22	14.89	46.71	37.32	89.26	45.85	53.96	35.01	98.97	80.14	76.77	99.74
Nutritional Status													
Calories: Deficient	90.89	27.70	5.50	39.02	17.28	75.93	15.31	24.15	9.05	96.68	65.37	43.80	99.38
Not Deficient	98.51	18.75	11.64	51.71	31.79	88.78	29.45	39.36	19.91	98.34	65.99	67.01	99.91
Protein: Deficient	88.21	28.87	3.38	43.42	15.03	68.71	10.01	18.39	4.84	96.11	59.84	37.44	99.31
Not Deficient	98.43	20.15	11.55	45.74	29.76	90.43	29.53	39.49	20.06	98.32	69.49	65.59	99.82
Vitamin A: Deficient	94.78	19.97	5.62	41.17	10.12	78.77	15.62	22.95	8.06	95.11	62.06	44.04	99.27
Not Deficient	94.02	26.70	10.58	47.92	35.61	84.37	26.90	38.02	19.07	99.41	68.71	63.16	99.92

Source: SUSENAS 1978 data tapes, Biro Pusat Statistik, Jakarta.

Table 6: PER CAPITA DAILY CONSUMPTION OF FOODS FOR HOUSEHOLDS REPORTING CONSUMPTION, INDONESIA, 1978
(in grams, except for eggs which are in units)

	Rice	Corn	Wheat	Cassava	Potatoes	Fish	Meat & Poultry	Eggs 1/	Dairy Products	Vegetables	Legumes	Fruits	Other
Indonesia	340.9	215.5	67.6	176.8	107.3	43.4	32.3	0.18	20.1	147.8	49.7	100.4	225.8
Region													
DKI Jakarta	319.4	43.7	20.4	50.0	34.1	30.8	27.4	0.25	23.0	101.7	66.0	90.9	257.2
West Java	408.3	87.4	23.5	129.9	69.6	29.2	31.1	0.19	20.6	138.1	47.9	94.2	238.0
Central Java	275.3	272.6	150.6	212.1	132.0	15.3	20.4	0.17	19.5	154.6	47.5	75.3	206.3
DI Yogyakarta	217.1	231.3	14.2	210.6	76.6	08.1	14.5	0.17	15.3	149.1	39.7	54.7	231.0
East Java	245.5	236.3	101.3	207.4	98.6	22.7	21.5	0.15	21.6	158.1	50.9	86.7	179.3
Sumatra	412.6	109.0	52.4	180.6	93.0	68.9	36.2	0.18	18.2	152.5	52.6	114.0	251.2
Bali & Nusatenggara	356.1	279.9	83.2	215.1	286.1	43.8	58.5	0.17	24.2	172.5	41.3	123.0	205.7
Kalimantan	419.7	104.3	44.8	155.0	77.6	94.4	38.4	0.17	19.1	143.5	55.2	120.2	291.4
Sulawesi	364.6	199.7	57.8	150.9	171.7	91.9	41.3	0.17	20.6	120.2	41.1	129.9	278.3
Maluku & Irian Jaya	233.7	60.5	55.2	258.7	136.1	122.3	29.6	0.21	22.6	147.4	44.4	138.5	234.0
Location													
Urban	324.2	91.7	35.2	78.4	46.0	49.5	27.3	0.21	23.0	136.4	63.6	97.2	237.3
Rural	344.5	224.6	78.4	190.2	125.4	42.2	34.8	0.16	17.7	150.1	46.1	101.2	223.4
Season													
February	329.8	204.6	84.3	180.1	116.4	44.6	34.2	0.18	20.5	160.9	58.5	119.8	246.2
May	354.2	257.6	52.4	176.6	102.3	44.2	33.4	0.18	19.6	145.8	48.7	97.1	219.4
November	337.8	193.4	53.2	174.3	104.0	41.5	30.1	0.17	20.3	136.2	42.1	84.3	210.9
Expenditure Class													
Lower 40%	310.7	261.6	110.1	215.2	157.7	34.2	32.2	0.16	24.6	143.4	41.2	88.1	181.0
Middle 30%	346.3	194.6	73.1	170.2	117.4	41.4	30.2	0.15	18.2	142.0	43.3	92.8	211.5
Upper 30%	369.3	155.2	49.8	139.5	76.6	54.8	33.1	0.20	20.2	158.8	62.1	113.0	295.5
Nutritional Status													
Calories: Deficient	260.8	225.8	61.5	122.8	78.0	25.9	19.0	0.13	16.3	115.6	36.1	58.7	147.8
Not Deficient	428.7	197.3	71.0	225.3	126.2	61.2	40.6	0.21	22.2	185.5	65.7	132.8	317.9
Protein: Deficient	246.8	235.3	74.3	172.8	101.3	17.0	13.9	0.11	15.3	107.8	63.9	53.9	137.3
Not Deficient	396.5	196.7	66.3	179.8	109.3	56.6	46.4	0.19	20.9	173.7	24.7	117.9	283.8
Vitamin A: Deficient	327.1	195.6	60.9	162.2	60.3	33.6	23.6	0.14	16.4	97.0	38.3	61.7	172.8
Not Deficient	352.7	228.1	70.6	187.4	118.6	51.2	36.6	0.19	21.5	189.1	58.5	123.3	270.5

1/ Eggs are given in units.

Source: SUSENAS 1978 data tapes, Biro Pusat Statistik, Jakarta.

capita daily consumption of cassava, a main staple, ranges from 47 grams in Jakarta to 209 grams in Central Java, while in the Outer Islands the range is much smaller, from 150 grams in Sulawesi to 258 grams in Maluku and Irian Jaya. Overall, consumption is higher in the Outer Islands, particularly in Bali, and Maluku and Irian Jaya. Of course Table 6 has to be interpreted in conjunction with Table 4 which shows the proportions of households reporting consumption of different food groups.

There is no significant difference between the urban and rural consumption of rice, the major staple, or of fish and "other" food items. However, of the households reporting consumption, the rural population consumes more of most other food groups, with the exception of milk, legumes, and eggs. Consumption of meat, while higher in rural areas of the Outer Islands, is consumed in similar quantities in urban and rural areas on Java. Finally, fruits are consumed in larger quantities in the urban areas of Java and in the rural areas of the Outer Islands.

For most foods, no pronounced seasonal variations in consumption are apparent. A very general pattern emerges for corn, wheat, fish and "other" foods: consumption is highest in February and lowest in November. Consumption of vegetables also appears to be at its lowest in November.

Clearly, households which are deficient in calories and protein consume less rice, by margins of about 40%, than those which are not. The former rely more on corn, but the difference in per capita consumption levels is small. Households poor in calories also report lower consumption of cassava and potatoes. Vitamin A deficiency appears to be associated with relatively low consumption levels of potatoes, fish, dairy products, vegetables, legumes and fruits.

2.2.2 Food Prices

The SUSENAS data were collected in both urban and rural areas in all the provinces of Indonesia. Moreover, the survey was conducted in four rounds during different seasons of the year. Variations in food prices across regions and seasons will be used in this paper to derive price elasticities of demand for foods and nutrients.

The implied price for each food group was calculated by dividing total consumption expenditure on each group by the total quantity of the group. The results are shown in Table 7.

Residents of Jakarta and Maluku and Irian Jaya face the highest prices for most food groups, while prices are lowest in Central Java and Bali. The exception to this is the "other" foods category which shows the highest average price in Central Java and moderate prices in Maluku and Irian Jaya.

Prices are generally higher in urban than in rural areas. There are no significant seasonal price variations for most food groups, although a few exhibit very general patterns. The price of cassava is highest in February, while the prices of corn, fish, meat and legumes are highest in November.

As would be expected, rich households pay higher prices for all foods. This is also reflected in the price differences between population groups with and without nutrient deficiencies. However, the calorie-deficient population pays a somewhat higher price for rice compared with that without the deficiency.

Table 7: PRICES OF FOODS FOR HOUSEHOLDS REPORTING CONSUMPTION, INDONESIA, 1978
(In Rupiah per kilogram except for eggs which are in Rupiah per unit)

	Rice	Corn	Wheat	Cassava	Potatoes	Fish	Meat & Poultry	Eggs 1/	Dairy Products	Vegetables	Legumes	Fruits	Other
Indonesia	141	68	122	28	70	361	941	38	736	106	178	109	445
Region													
DKI Jakarta	156	105	146	45	129	620	1123	51	862	258	204	206	767
West Java	140	73	148	24	70	401	978	43	702	92	141	120	369
Central Java	136	53	89	21	43	293	966	33	730	66	157	88	778
DI Yogyakarta	139	43	114	22	58	363	948	31	762	80	143	107	560
East Java	137	65	91	25	50	341	939	32	826	65	158	99	317
Sumatra	152	72	139	30	100	371	1015	43	648	161	202	113	361
Bali & Nusatenggara	132	69	145	36	60	337	559	18	839	107	217	103	339
Kalimantan	147	90	162	32	111	384	978	52	737	107	276	125	427
Sulawesi	144	63	130	51	76	328	750	41	642	125	238	101	300
Maluku & Irian Jaya	157	67	163	54	145	343	1159	62	679	154	310	149	464
Location													
Urban	153	82	142	36	111	462	1071	45	759	151	198	157	552
Rural	139	65	118	27	62	341	914	36	731	96	174	100	424
Season													
February	148	67	124	31	70	341	913	38	714	104	168	107	541
May	128	62	116	27	70	355	908	37	750	106	177	112	405
November	149	76	126	26	69	388	1007	38	744	106	189	109	387
Expenditure Class													
Lower 40%	139	63	115	27	59	324	902	34	738	91	166	96	335
Middle 30%	141	68	121	27	67	349	936	38	733	105	176	107	375
Upper 30%	149	74	131	31	86	419	994	42	737	125	194	129	651
Nutritional Status													
Calories: Deficient	144	66	114	27	63	349	944	36	749	96	169	108	535
Not Deficient	139	70	131	30	77	376	936	40	721	117	188	111	338
Protein: Deficient	138	64	110	26	58	348	939	35	754	91	166	102	565
Not Deficient	143	70	129	30	77	370	941	40	724	115	185	114	366
Vitamin A: Deficient	138	67	119	29	68	353	938	37	741	109	172	109	423
Not Deficient	144	69	124	28	71	368	943	39	732	102	183	109	464

1/ The price of eggs is given in Rupiah per egg.

Source: SUSENAS 1978 data tapes, Biro Pusat Statistik, Jakarta.

2.2.3 Consumption of Calories and Other Nutrients

Per capita daily consumption of different nutrients is shown in Table 8 which essentially summarizes all the differences shown in previous tables. [1] Consumption of nutrients is generally higher in the Outer Islands than in Java. The residents of Sumatra and Kalimantan show especially high intakes of nutrients, while the residents of Central Java and Yogyakarta have relatively lower intakes. The population of Sulawesi apparently consumes less calcium and vitamin C than elsewhere.

Overall, the consumption of nutrients does not vary significantly between urban and rural areas. Consumption of fat is slightly higher in urban areas, while the consumption of calories, carbohydrates and vitamin C is higher in rural areas. Protein and vitamin A consumptions are higher in the urban areas than in rural areas of Java; the reverse holds in the Outer Islands.

There are no significant seasonal variations in the intakes of different nutrients.

2.2.4 Deficiencies in Calories and Other Nutrients

The estimated per capita daily consumption of nutrients given in Tables 8 can be compared with estimated minimal requirements reported in

---

[1] The reader should note that these consumption levels are derived from information on purchased foods. The impact of the level of income/expenditures is evident across the board. The higher the levels of household expenditures, the higher the level of consumption of nutrients. The figures suggest that households deficient in calories and protein are also likely to be deficient in other nutrients. The differences between purchases and actual intakes may be substantial for a variety of reasons.

Table 8: PER CAPITA DAILY CONSUMPTION OF NUTRIENTS, INDONESIA, 1978

	Calories (calories)	Protein (grams)	Fat (grams)	Carbohydrates (grams)	Calcium (mg.)	Iron (mg.)	Vitamin A (int. unit)	Thiamine (mg.)	Riboflavin (mg.)	Niacin (mg.)	Vitamin C (mg.)
Indonesia	1987	50.94	30.59	381	290	9.73	5687	0.89	0.70	14.7	152
Region											
DKI Jakarta	1946	56.03	45.31	333	335	9.21	4967	0.93	0.70	13.5	136
West Java	2097	53.97	26.13	409	257	8.65	3770	0.86	0.67	16.2	138
Central Java	1605	38.53	21.60	321	262	8.57	5346	0.77	0.62	10.7	135
DI Yogyakarta	1584	35.09	25.56	312	293	9.26	3691	0.78	0.57	9.4	125
East Java	1664	41.81	26.01	326	299	9.61	4783	0.84	0.67	11.5	122
Sumatra	2408	62.20	42.75	445	333	11.55	8244	1.01	0.82	18.3	242
Bali & Nusatenggara	2229	55.44	33.22	430	287	11.18	11241	0.77	0.77	16.7	153
Kalimantan	2431	67.30	36.88	456	348	11.16	5831	1.08	0.76	19.6	151
Sulawesi	2253	62.21	37.93	419	252	9.98	5041	0.91	0.69	18.5	117
Maluku & Irian Jaya	2010	57.14	43.23	352	347	10.51	8184	0.86	0.78	15.2	221
Location											
Urban	1912	53.91	37.07	344	308	9.44	5361	0.90	0.70	14.0	141
Rural	2002	50.34	29.27	388	286	9.79	5754	0.89	0.70	14.9	154
Season											
May	1990	52.10	30.85	380	291	9.94	5720	0.91	0.72	15.0	155
November	1983	49.71	30.29	381	288	9.50	5652	0.86	0.67	14.5	149
Expenditure Class											
Lower 40%	1747	41.97	23.20	346	254	8.56	5367	0.76	0.62	12.6	146
Middle 30%	1988	49.95	28.63	385	274	9.58	5337	0.88	0.67	14.6	146
Upper 30%	2279	62.90	41.54	418	349	11.30	6423	1.05	0.81	17.4	165
Nutritional Status											
Calories: Deficient	1406	35.67	19.49	275	203	6.86	3734	0.64	0.52	10.4	102
Not Deficient	2697	69.62	44.14	509	395	13.23	8074	1.19	0.92	20.0	212
Protein: Deficient	1356	29.54	16.47	275	187	6.34	3643	0.56	0.46	9.2	98
Not Deficient	2408	65.24	40.01	451	358	11.99	7052	1.11	0.35	18.1	188
Vitamin A: Deficient	1721	41.92	38.27	338	198	7.11	1648	0.70	0.50	12.3	69
Not Deficient	2238	59.49	22.47	421	376	12.20	9508	1.07	0.87	17.0	230

Source: SUSENAS 1978 data tapes, Biro Pusat Statistik, Jakarta.

- 20 -

Table 9. These requirements have been set for three broad age categories,
namely individuals under 5, between 5 and 10, and 10 and over, according to
tables calculated for the Indonesian population. 1/ The average requirements
were set for each age group. An additional 15% of the average was added to
the minimum to allow for pregnant and lactating women. For calories, the
minimal requirements are as follows:

Age Group	Minimal Daily Caloric Requirement
Under 5	1166
5-9	1654
10 and over	2126

The minimal per capita daily requirement of calories for a region,
for example, would be the average requirement which takes into account the
distribution of the population among the three age categories listed above.
Thus a region with a relatively high proportion of children would have a
relatively low minimal average requirement. Table 9 gives the estimated
minimal per capita daily requirements of nutrients by region. We can see that
the provinces of Java, which have higher proportions of the population in the
age group 10 and over compared with the Outer Islands, also have higher
average requirements of calories, as would be expected.

1/   From Djumpaolias et. al., National Workshop on Food and Nutrition, Bogor,
10-14 July, 1978.

Table 9: ESTIMATED MINIMAL PER CAPITA DAILY REQUIREMENTS OF NUTRIENTS, INDONESIA

	Calories (calories)	Protein (grams)	Calcium (mg.)	Iron (mg.)	Vitamin A (int. units)	Thiamine (mg.)	Riboflavin (mg.)	Niacin (mg.)	Vitamin C (mg.)
Indonesia	1933	40.43	0.56	13.76	3291	0.78	1.10	12.84	28.55
Region									
DKI Jakarta	1926	40.35	0.56	13.77	3286	0.78	1.10	12.80	28.60
West Java	1935	40.51	0.56	13.79	3300	0.79	1.10	12.86	28.62
Central Java	1937	40.56	0.56	13.79	3305	0.79	1.10	12.87	28.64
DI Yogyakarta	1978	41.60	0.57	14.06	3413	0.80	1.13	13.15	29.24
East Java	1958	41.04	0.56	13.90	3354	0.79	1.12	13.01	28.88
Sumatra	1909	39.82	0.56	13.59	3227	0.78	1.09	12.69	28.19
Bali & Nusatenggara	1917	40.01	0.56	13.64	3147	0.78	1.09	12.74	28.30
Kalimantan	1914	39.94	0.56	13.63	3240	0.78	1.09	12.72	28.26
Sulawesi	1907	39.76	0.56	13.58	3220	0.78	1.09	12.67	28.15
Maluku & Irian Jaya	1913	39.97	0.56	13.66	3246	0.78	1.09	12.71	28.34
Location									
Urban	1929	40.41	0.56	13.78	3291	0.78	1.10	12.82	28.60
Rural	1939	40.43	0.56	13.75	3291	0.79	1.10	12.84	28.54
Season									
February	1930	40.39	0.56	13.75	3287	0.78	1.10	12.83	28.55
May	1935	40.48	0.56	13.76	3296	0.79	1.11	12.86	28.57
November	1932	40.41	0.56	13.75	3289	0.79	1.10	12.84	28.54
Expenditure Class									
Lower 40%	1961	41.19	0.57	13.97	3372	0.80	1.12	13.04	29.03
Middle 30%	1913	39.91	0.56	13.62	3226	0.78	1.09	12.71	28.23
Upper 30%	1916	39.97	0.56	13.63	3242	0.78	1.09	12.73	28.26
Nutritional Status									
Calories: Deficient	1932	40.40	0.56	13.74	3287	0.79	1.10	12.84	28.51
Not Deficient	1933	40.46	0.56	13.77	3296	0.79	1.10	12.85	28.60
Protein: Deficient	1938	40.54	0.56	13.77	3301	0.79	1.11	12.88	28.60
Not Deficient	1929	40.36	0.56	13.74	3284	0.78	1.10	12.82	28.52
Vitamin A: Deficient	1936	40.51	0.56	13.77	3299	0.79	1.11	12.87	28.59
Not Deficient	1929	40.36	0.56	13.74	3284	0.78	1.10	12.82	28.53

Source: SUSENAS 1978 data tapes, Biro Pusat Statistik, Jakarta.

To continue to discuss caloric requirements, deficiencies in the consumption of calories can come about because the average consumption falls short of the average requirement and, in addition, because whatever calories are available are not distributed according to requirements. The first says that there are not enough total calories to satisfy everyone's minimal requirements. The second says that, even if there are enough calories in the aggregate, because some people are consuming more than their minimal requirements, there are others who do not consume enough to satisfy their requirements. Comparing Tables 8 and 9, the estimated total availability of calories in Indonesia comes to 104% of what would be required to meet everyone's needs.[1]/ However, when the comparison is made on a regional basis, we see that there is a problem of aggregate availability of calories in all the provinces of Java, as well as a slight shortfall in Maluku and Irian Jaya, whereas there are more calories than the estimated requirements in all the other regions in the Outer Islands.

This suggests that there may be quite a serious problem of caloric deficiency in Java. In Central Java and Yogyakarta, the available calories come to only 82% and 77% of what would be required. If those in Yogyakarta who do not have a deficiency consume exactly their minimal requirement of calories and no more, while those with a deficiency consume on average 1,300 calories (87% of the average), then 70% of the population would have a deficiency in calories.

The optimistic aggregate picture in the Outer Islands does not imply that there would be no deficiencies in calories there. As pointed out earlier, distributional problems would mean deficiencies for some. The data

---

1/   This compares with an estimate of daily per capita calorie supply of 2,272 calories in 1977, or 102% of requirement, given in the World Development Report 1982.

for the Outer Islands are aggregated into major islands; there could be a great deal of variation among the provinces within each one.

Estimates of the proportion of the population with deficiencies in different nutrients are given in Table 10. These estimates are obtained on a case-by-case basis for all the sample households so that the household's food consumption and the implied nutrient consumption are measured against the household's own requirements, given its age distribution. Note that an implicit assumption is made that there are no distributional problems within the household; that is, if the household total requirement for a nutrient is met, then all the household members are considered to have satisfied their requirements.

2.2.5 Profiles of Populations Deficient in Calories, Protein and Vitamin A

The data presented thus far show considerable variations in food and nutrition consumption patterns across regions, locations (urban/rural) and by income class. The next step requires an explicit association between consumption patterns, market conditions and household characteristics. In this section we address the basic characteristics distinguishing households which meet minimal requirements in calories, protein and vitamin A, and those which do not. Such a distinction, combined with the regional differences discussed before, may help in the formulation of policies aimed at target groups defined according to some broad population characteristics.

Tables 11, 12, and 13 give a comparison of the characteristics of households meeting calorie, protein and vitamin A requirements, and households which do not. In both Java and the Outer Islands, the most significant discriminating factor between the two groups is household monthly expenditures. Moreover, households with deficiencies are larger than those without them.

Table 10: PROPORTIONS OF THE POPULATION WITH NUTRIENT DEFICIENCIES, INDONESIA, 1978
(%)

	Calories	Protein	Calcium	Iron	Vitamin A	Thiamine	Riboflavin	Niacin	Vitamin C
Indonesia	54.32	39.75	0.06	81.82	45.92	48.97	86.49	47.86	11.80
Region									
DKI Jakarta	55.76	28.16	0.03	86.72	41.46	44.60	86.57	52.89	16.25
West Java	47.59	30.55	0.00	87.49	56.61	49.28	88.16	33.57	15.84
Central Java	74.29	63.46	0.00	85.67	47.86	60.68	89.93	73.01	10.92
DI Yogyakarta	81.46	74.20	0.00	87.87	54.20	62.69	95.71	89.46	10.57
East Java	72.21	56.81	0.17	84.19	53.31	54.59	89.97	70.58	11.12
Sumatra	29.80	17.06	0.06	71.94	24.58	34.13	78.91	22.22	04.50
Bali & Nusatenggara	44.55	36.58	0.00	74.85	34.50	41.10	81.47	38.95	16.05
Kalimantan	39.93	14.14	0.00	72.52	40.51	39.60	82.39	18.10	09.08
Sulawesi	41.55	22.87	0.00	79.59	52.27	47.06	84.63	28.14	19.66
Maluku & Irian Jaya	53.91	32.96	0.43	79.19	31.07	53.10	80.58	43.60	03.37
Location									
Urban	57.90	33.05	0.11	83.23	39.47	47.25	85.26	49.61	12.17
Rural	53.59	41.09	0.05	81.54	47.21	49.31	86.74	47.50	11.72
Season									
February	52.98	39.18	0.05	77.66	40.64	46.05	83.35	47.78	11.80
May	55.63	38.93	0.11	82.71	47.69	50.13	86.61	46.77	13.46
November	54.31	41.22	0.01	85.26	49.59	50.80	89.69	49.10	10.02
Expenditure Class									
Lower 40%	68.83	58.09	0.11	87.77	52.49	63.87	90.88	62.91	15.79
Middle 30%	53.04	36.71	0.01	84.16	48.75	49.11	88.33	45.22	11.48
Upper 30%	37.37	19.27	0.04	72.08	34.90	30.15	79.21	31.57	07.11

Source: SUSENAS 1978 data tapes, Biro Pusat Statistik, Jakarta.

Table 11: BASIC CHARACTERISTICS OF HOUSEHOLDS DEFICIENT AND NOT DEFICIENT IN CALORIES, INDONESIA, 1978

Household Characteristic	JAVA			OUTER ISLANDS		
	Deficient	Not Deficient	t-statistic for difference	Deficient	Not Deficient	t-statistic for difference
Household Monthly Expenditure (Rp)	26931	57214	-12.9	27771	40039	-14.1
Number of Members of Age						
0-4	0.804	0.713	2.3	1.201	0.888	9.6
5-9	0.583	0.495	2.8	0.781	0.628	5.5
10 +	3.610	3.389	2.8	3.912	3.266	11.0
% of Households with Education of Head of Household						
Elementary School	0.507	0.478	1.3	0.563	0.582	-1.2
Junior High School	0.070	0.091	-1.8	0.095	0.103	-0.9
Senior High School	0.046	0.134	-7.7	0.083	0.088	-0.6
Higher Education	0.008	0.038	-5.3	0.020	0.016	0.8
% of Households with Some Income from						
Agriculture	0.516	0.485	1.4	0.482	0.608	-8.0
Industry	0.169	0.134	2.3	0.094	0.073	2.4
Services	0.529	0.507	1.0	0.460	0.363	6.2
Government	0.094	0.174	-5.8	0.166	0.151	1.3
Transfers	0.931	0.948	-1.6	0.860	0.882	-2.1
% of Households Urban	0.575	0.527	2.3	0.410	0.622	-13.5

Source: Data tapes of SUSENAS 1978, May round, Biro Pusat Statistik, Jakarta.

Table 12: BASIC CHARACTERISTICS OF HOUSEHOLDS DEFICIENT AND NOT DEFICIENT IN PROTEIN, INDONESIA, 1978

Household Characteristic	JAVA			OUTER ISLANDS		
	Deficient	Not Deficient	t-statistic for difference	Deficient	Not Deficient	t-statistic for difference
Household Monthly Expenditure (Rp)	21846	50416	-12.8	25022	37318	-11.3
Number of Members of Age						
0-4	0.748	0.794	-1.2	1.075	1.006	1.7
5-9	0.582	0.529	1.7	0.871	0.648	6.5
10 +	3.615	3.466	2.0	4.342	3.338	13.9
% of Households with Education of Head of Household						
Elementary School	0.500	0.495	0.2	0.574	0.574	0.01
Junior High School	0.051	0.100	-4.4	0.071	0.107	-3.0
Senior High School	0.028	0.117	-8.3	0.043	0.097	-4.8
Higher Education	0.002	0.032	-5.5	0.009	0.020	-2.1
% of Households with Some Income from						
Agriculture	0.576	0.446	6.4	0.574	0.550	1.2
Industry	0.183	0.136	3.2	0.110	0.074	3.3
Services	0.498	0.543	-2.2	0.468	0.388	4.1
Government	0.067	0.167	-7.6	0.085	0.175	-6.3
Transfers	0.934	0.939	-0.5	0.849	0.879	-2.2
% of Households Urban	0.632	0.497	6.7	0.528	0.533	-0.2

Source: Data tapes of SUSENAS 1978, May round, Biro Pusat Statistik, Jakarta.

Table 13: BASIC CHARACTERISTICS OF HOUSEHOLDS DEFICIENT AND NOT DEFICIENT IN VITAMIN A,
INDONESIA, 1978

Household Characteristic	JAVA			OUTER ISLANDS		
	Deficient	Not Deficient	t-statistic for difference	Deficient	Not Deficient	t-statistic for difference
Household Monthly Expenditure (Rp)	27243	47987	-9.2	27502	39651	-13.8
Number of Members of Age						
0-4	0.766	0.781	-0.4	1.115	0.958	4.7
5-9	0.588	0.516	2.3	0.703	0.686	0.6
10 +	3.651	3.411	3.2	3.642	3.471	2.8
% of Households with Education of Head of Household						
Elementary School	0.520	0.473	2.3	0.569	0.577	-0.5
Junior High School	0.068	0.088	-1.8	0.079	0.113	-3.5
Senior High School	0.038	0.117	-7.3	0.072	0.095	-2.4
Higher Education	0.003	0.034	-5.6	0.016	0.019	-0.6
% of Households with Some Income from						
Agriculture	0.541	0.468	3.6	0.574	0.542	2.0
Industry	0.178	0.135	2.9	0.083	0.081	0.2
Services	0.516	0.528	-0.6	0.397	0.409	-0.8
Government	0.085	0.160	-5.7	0.155	0.159	-0.4
Transfers	0.934	0.940	-0.6	0.875	0.972	0.3
% of Household Urban	0.595	0.520	3.7	0.490	0.560	-4.33

Source: Data tapes of SUSENAS 1978, May round, Biro Pusat Statistik, Jakarta.

The expenditure and household size differences between Java and the Outer Islands suggest that in the Outer Islands the presence of deficicies, in calorie consumption in particular, may be more related to poverty than in Java.

Education is a discriminating factor between the calorie-deficient and nondeficient populations in Java, but not in the Outer Islands. In the latter region, education is more significant in explaining the consumption of protein and vitamin A. The finding of the relative insignificance of education in the consumption of calories in the Outer Islands supports the notion that in these islands poverty may be the key issue, while in Java other factors may be important as well.

Households are also classified by their sources of income. In general, the likelihood of households with incomes from industry and agriculture failing to meet minimal nutritional requirements is smaller than those with incomes from other sources. Surprisingly, we find that a household reporting income from government employment is more likely to be deficient in each of the nutrients compared with a household not reporting any income from this source. A major difference between Java and the Outer Islands is that in the Outer Islands a household reporting income from agriculture is more likely to be deficient in calories than otherwise; the opposite holds for Java.

This last point is probably reflected in the rural/urban differentials. Rural households are more likely to be better-off in terms of nutrition in Java, and worse off in terms of vitamin A in the Outer Islands.

Next, we turn to the determination of the specific causes of variations in food consumption and nutrition. This may help in understanding how households have become deficient in nutrients and in selecting policy instruments to alleviate the situation.

III.   CONCEPTUAL AND ANALYTICAL CONSIDERATIONS

In this section, we outline some basic theoretical and practical considerations which will guide us in dealing further with the data described in the previous section. Our basic objective here is to estimate marginal propensities to consume and demand elasticities for various foods and nutrients. These elasticities are the best available indicators of how households may respond to policies which change relative prices and the level and distribution of income. In addition, we wish to examine the effects of other variables, primarily household size and composition, education of the spouse of the household head (presumed to be the homemaker), as well as sources of income. These variables are believed to be critical in the determination of food and nutrition consumption but are not readily amenable to policy interventions. [1]/

We assume that households operate in a competitive food market and that "foods" constitute a (weakly) separable branch of the household's utility function which is identical for all households. This particular branch can be expressed either as a function of a vector of n food items, F_i, denoting both type and quantity,

$$u = u(F_1, F_2, \ldots\ldots, F_n) \tag{1}$$

or as a function of m nutrients (m < n) which are a linear transformation of foods through a matrix containing n x m coefficients α_{ij} (i=1, ...,n; j=1,....., m)

---

[1]/   On the relevance and use of elasticity estimates for policy options, see Reutlinger and Selowsky (1978).

which transform foods into nutrients. That is:

$$u' = u'(N_1, N_2, \ldots, N_m). \quad \underline{1/}$$ (2)

This assumption permits us to treat food or nutrient consumption as
if the household preallocates a particular budget to foods before entering the
market and maximizes its utility subject to this budget preallocation.
Furthermore, we assume that the household does not increase or decrease consumption of particular foods and nutrients with changes in the consumption of
other "nonfood" goods and services, unless these changes affect its budget
allocation to food. This general approach is supported by the relatively high
shares of food in the household budget.

Hence, we may a priori consider the family's relevant budget con-
straint as its expenditures on food (E_f). We thus have a basic demand system
with n equations of the type:

$$F_i = f_i(P_1, P_2, \ldots, P_n, E_f, R)$$ (3)

where $(P_1, P_2, \ldots, P_n)$ is a vector of food prices and R is a set of
relevant control variables which are discussed below. The implied "demand"
for nutrients is thus

$$N_j = g_j(P_1, P_2, \ldots, P_n, E_f, R).$$ (4)

1/   We do not explicitly discuss here the different approaches possible. We
might also consider that equation (1) is derived from equation (2). That is,
we assume that a sophisticated consumer's demand for foods is derived from his
demand for a particular diet. See Lancaster (1966) for a theoretical discussion and Chernichovsky (1977) for an application of this. The arguments in
(1) and (2) can be put in per capita or per adult-equivalent terms.

As the demand for (and consumption of) nutrients is the prime concern of this paper, we shall focus on a few important parameters. The implicit demand for each nutrient, N_j, $(j = 1, 2, . . ., m)$, is a linear tranformation of the entire foods vector:

$$N_j = \sum_{i=1}^{n} \alpha_{ij} F_i$$

where α_{ij} is a coefficient transforming a particular quantity of food F_i into a certain quantity of nutrient, N_j. This tranformation implies that the sensitivity of demand for a particular nutrient to a change in income or expenditures, its income or expenditure elasticity η, is

$$\eta_{N_j E} = \sum_{i=1}^{n} \frac{\alpha_{ij} F_i}{N_j} \eta_{F_i E} \qquad (6)$$

Correspondingly, the relevant price elasticity of demand with respect to a particular price, say P_1, is

$$\eta_{N_j P_1} = \sum_{i=1}^{n} \frac{\alpha_{ij} F_i}{N_j} \eta_{F_i P_1} \qquad (7)$$

Equations (5) and (6) indicate that the impact of changes in income and price on the consumption of nutrient N_j is a function of the share of each particular food item F_i in the total consumption of this nutrient $(k_{ij} = \frac{\alpha_{ij} F_i}{N_j})$ and of the income and price elasticities of that food.

Relationships (6) and (7) are of particular interest. They signify that the effect of a change in income or in price P_i on the consumption of a particular nutrient is a function of the impact of the change on the entire food basket. It is not at all clear, a priori, how a particular change will affect the consumption of particular nutrients, unless we know the income and price elasticities for each food as well as its share in the consumption of each nutrient, k_{ij}.

It is thus clear that, with a change in incomes or relative prices, the resulting change in consumption of foods with low income or price elasticities may have a significant impact on the consumption of a particular nutrient, N_j, if these foods are significant sources of that nutrient. By the same token, the resulting change in the consumption of foods with a low contribution to N_j may also have a significant impact on the consumption of N_j, if these foods have high income or price elasticities of demand.

These general considerations emphasize the significance of our empirical analysis, which must deal with both income and price elasticities, as well as the shares of particular foods in the consumption of nutrients (k_{ij}).

Next we turn to some basic empirical and computational considerations. First, we deal with the definitions of the relevant variables. As indicated in the previous section, foods have been grouped into 13 major categories. In relation to this grouping, prices present a serious conceptual and empirical problem. As price information is not directly available in the data, which give only the total expenditure on, and the quantity of, each food item, prices have had to be defined as the average expenditure per item for the group $(P_i = E_i/F_i)$. This average implicit price does not necessarily represent the marginal price that consumers face, but is the only information available.

For those households reporting no expenditure at all on a particular item, a local average price was assigned on the assumption that each household faces an average price common to its area at any given time. [1]

Hence any given price (P_i) is defined as the total expenditure on a particular group consisting of k items, divided by total quantities, measured in homogeneous units, in this group. Hence,

$$P_i = \frac{\sum_{\ell=1}^{k} E_\ell}{\sum_{\ell=1}^{k} Q_\ell} = \sum_{\ell=1}^{k} P_\ell \frac{Q_\ell}{\sum_{\ell=1}^{k} Q_\ell} \tag{8}$$

which is a weighted average price per kilogram of a group of foods. The weights are the quantity shares of each food in the group. The price thus becomes a function not only of the average individual prices of foods, but also of the relative quantities consumed within each food category. As a result, households buying more expensive items face higher group prices.

The discussion of prices leads to Theil's notion of a quality-quantity trade-off that may be particularly relevant here (Theil 1952). We assume that items within each group are close substitutes, and that people

---

[1] To make the computation practical for each region (defined previously), we estimated separately for Java and the Outer Islands:

$$P_g = P_g^o + \sum_{i=1}^{K} \alpha_i d_i + \beta d_u + \sum_{k=1}^{M} \gamma_k d_k + \nu$$

where P_g is the average price of a group; d_i are dummy variables representing K subregions; d_u is a dummy variable representing urban location; and d_k are dummy variables representing the different months.

prefer one item to another because of various qualitative differences, manifested in prices. As wealthier households tend to buy more expensive foods, the group price becomes a function of household income and other household characteristics such as size, composition, etc. [1]/

From the relationships discussed thus far, we can also derive an implicit price of a nutrient. When foods are measured in identical units, say kilograms, and each food (group) has a price P_g per kilogram, one can define the following:

$$P_{N_j} = \sum_{i=1}^{N} P_i k_{ij}$$

P_{N_j} is thus the price of nutrient j coming from a kilogram of a particular basket of food. This price is seen to be a function of all previous variables: the relative contribution of various foods to any particular nutrient; the particular mix of food groups in any given diet; the particular composition of any food group; and of course food prices. This measure (P_{N_j}) can be used as a general indicator of efficiency in consumption with regard to different nutrients, as it conveys the effects of compositional changes in the food basket on costs of nutrients.

In this paper we focus on the quantitative aspects of consumption which lead to actual consumption of nutrients. It should be noted that prices for relatively highly aggregated groups, particularly meat and poultry, and fruits and vegetables, may vary systematically with income/expenditures.

---

[1]/ This general problem may already exist in the case of individual food items; wealthier households may be paying higher average prices for them.

For empirical purposes, we use household monthly expenditures, instead of just expenditures on food as advocated earlier by conceptual considerations. The two variables are highly correlated and thus make almost no difference from a statistical viewpoint. Furthermore, household expenditures are commonly accepted as a good proxy for permanent income and come closer to any operational concept of income.

Household size and composition enter the demand equation through the numbers in three age groups: children 0-4, children 5-9 and members ages 10 and above.

$$Ln(F_i) = \beta_o + \beta_e Ln(E) + \sum_{i=1}^{13} \beta_i Ln(P_i) + \sum_{s=1}^{3} \gamma_s S_s + \sum_{k=1}^{4} \delta_k D_k + \mu_i \quad (9)$$

where F_i = quantity of food of type i purchased

E = household monthly expenditure

P_i = price of commodity i

S_s = number of household members in age group s

D_k = dummy variable for education/source of income

An alternative reduced form of this function would replace F_i by nutrients N_j (j = 1,..., m).

Finally, we discuss the choice of functional form and the actual estimation procedure. We have chosen the double-logarithmic function in expenditures and prices. While the proposed function does not meet the general restrictions applying to demand equations, its practical appeal and

its descriptive properties are overriding. [1] Similar analyses, (Alderman and
Timmer, 1980; Chernichovsky, 1977) point to the goodness of fit of this function. Food and nutrition consumption, as well as incomes, are known to have
skewed distributions; the logarithmic transformation brings them closer to
normality. Furthermore, since we are dealing with a full range of foods and
particularly of nutrients, it is efficient to obtain direct elasticities which
are independent of the units of measurements of quantities. The problem with
this function of forcing constant elasticities over the entire income/expen-
diture or price range is mitigated to some extent by confining part of the
discussion to the population deficient in calories, broken down by Java and
the Outer Islands.

IV.   SOURCES OF VARIATIONS IN THE DEMAND FOR FOODS AND NUTRIENTS

In this section, we report and discuss the estimated income and price
elasticities of demand for foods and nutrients, as well as the effect of
household characteristics on consumption patterns. The analysis is done
separately for Java and the Outer Islands, and for households which are
deficient in calories.

4.1   Sources of Calories and Other Nutrients

As suggested earlier by the conceptual framework, a clearer understanding of how changes in incomes and relative prices affect the consumption
of nutrients requires an appreciation of the contributions of different foods
to the various nutrients. These are given for the lower 40% and upper 30%

---

[1] Households with zero expenditure on any food item are assigned a value of
.001 in the regression analysis.

expenditure classes for Java and the Outer Islands in Tables 14 and 15.

Calories come primarily from rice, which also contributes substantially to protein, carbohydrates, thiamine, iron, riboflavin, niacin and, to a lesser extent, fat. Cassava and corn are also important contributors of calories, especially among the poor in Java. Fish, legumes, fruits and vegetables are other important nonrice sources. It is important to note that the dependence on rice as a source of calories increases with income.

Next to rice, fish is an important source of protein, especially in the Outer Islands. Other sources are vegetables and legumes. It is clear from the tables that, as people get wealthier, they tend to get a larger proportion of their protein from meat and poultry.

Fat comes in relatively substantial amounts from legumes and "other" sources, which include various fish products, oils, and coconut. Again the wealthier households tend to get a higher proportion of fat from "other" products and meat, and less from rice and vegetables.

Carbohydrates come primarily from rice and cassava, as well as other food products. To a lesser degree this is the case for corn among the poor.

For the poor, calcium comes primarily from cassava, vegetables and legumes. Fish plays a major role as a source of calcium in the Outer Islands. Similarly, iron comes primarily from vegetables, legumes and cassava, especially in Java and among the poor. Fish is also important as a source of iron in the Outer Islands.

Vitamin A comes primarily from fruits and vegetables, as well as corn and potatoes. For the poor, corn is relatively more important, while for the better-off, it is potatoes and fruits.

Table 14: SOURCES OF NUTRIENTS, JAVA, INDONESIA, 1978
(%)

Expenditure Group	Calories	Protein	Fat	Carbohydrates	Calcium	Iron	Vitamin A	Thiamine	Riboflavin	Niacin	Vitamin C
Rice - Total	57.30	44.92	13.16	64.27	9.46	30.29	0.00	45.00	25.92	53.51	0.00
Lower 40%	52.95	43.21	15.45	57.68	9.53	28.70	0.00	42.66	23.74	48.88	0.00
Upper 30%	58.67	43.28	9.25	68.84	8.31	29.63	0.00	44.39	26.10	55.59	0.00
Corn - Total	5.13	5.58	4.61	6.50	1.55	5.18	8.43	6.25	5.81	6.74	2.98
Lower 40%	8.29	8.90	7.52	10.59	2.44	7.66	11.05	9.06	8.87	10.84	4.84
Upper 30%	2.26	2.42	1.88	2.76	0.65	2.69	5.32	3.56	2.67	2.73	0.93
Wheat - Total	0.91	0.90	0.76	0.97	0.40	1.02	0.53	1.41	0.74	0.87	0.00
Lower 40%	1.27	1.19	1.21	1.32	0.49	1.28	0.70	1.67	0.95	1.09	0.00
Upper 30%	0.66	0.73	0.38	0.75	0.34	0.90	0.25	1.37	0.61	0.79	0.00
Cassava - Total	9.16	3.06	2.08	10.47	10.94	8.06	0.00	5.60	1.36	4.25	11.26
Lower 40%	12.96	4.58	3.29	14.42	14.04	11.03	0.00	7.50	1.75	5.90	10.96
Upper 30%	5.04	1.50	0.81	6.11	6.86	4.63	0.00	3.38	0.84	2.41	10.07
Potatoes - Total	1.12	0.69	0.55	1.30	1.74	1.40	7.14	1.68	1.28	1.02	2.88
Lower 40%	1.10	0.69	0.65	1.22	1.61	1.27	5.62	1.52	1.13	0.90	2.29
Upper 30%	1.06	0.63	0.37	1.31	1.61	1.40	8.06	1.74	1.30	1.14	3.26
Fish - Total	1.67	11.53	2.38	0.00	10.78	4.42	0.96	0.65	4.31	6.70	0.00
Lower 40%	1.65	11.46	2.31	0.00	10.19	4.20	0.88	0.51	4.01	6.23	0.00
Upper 30%	1.71	11.25	2.34	0.00	11.02	4.57	1.13	0.82	4.49	7.30	0.00
Meat & Poultry - Total	0.43	1.44	1.52	0.00	0.15	0.82	0.22	0.40	0.91	1.56	0.00
Lower 40%	0.09	0.34	0.40	0.00	0.03	0.17	0.05	0.09	0.19	0.31	0.00
Upper 30%	0.99	3.27	3.33	0.00	0.33	1.96	0.48	0.94	2.11	3.62	0.00
Eggs - Total	0.19	0.53	0.86	0.00	0.43	0.53	0.96	0.26	0.96	0.03	0.00
Lower 40%	0.08	0.22	0.45	0.00	0.19	0.21	0.46	0.11	0.36	0.01	0.00
Upper 30%	0.40	1.06	1.54	0.00	0.85	1.11	1.81	0.53	2.00	0.07	0.00
Dairy Products - Total	0.19	0.49	0.36	0.11	2.30	0.05	0.46	0.25	1.53	0.05	0.10
Lower 40%	0.01	0.03	0.05	0.00	0.14	0.00	0.06	0.01	0.09	0.00	0.05
Upper 30%	0.55	1.39	0.93	0.33	6.49	0.14	1.20	0.72	4.35	0.15	0.20
Vegetables - Total	4.36	9.49	3.48	4.89	37.79	25.17	71.97	17.17	44.25	15.73	72.27
Lower 40%	5.28	11.93	4.94	5.65	42.01	26.55	73.92	19.57	49.67	18.95	74.35
Upper 30%	3.61	7.20	1.99	4.39	32.23	23.99	68.67	15.17	37.15	12.42	70.27
Legumes - Total	4.27	14.48	15.61	2.08	18.30	13.60	0.67	18.07	8.59	5.76	0.29
Lower 40%	3.61	12.21	13.90	1.74	13.96	10.67	0.60	14.96	6.35	4.39	0.11
Upper 30%	5.34	17.74	17.43	2.63	24.12	17.72	0.74	22.48	11.78	7.88	0.58
Fruits - Total	1.06	0.54	0.48	1.40	1.65	2.20	6.57	1.28	1.87	1.18	7.97
Lower 40%	0.66	0.35	0.28	0.85	1.00	1.30	4.18	0.83	1.14	0.71	4.70
Upper 30%	1.66	0.83	0.79	2.25	2.58	3.55	10.26	1.99	2.94	1.85	12.82
Other - Total	14.16	6.28	54.06	7.97	4.44	7.21	0.94	1.92	2.39	2.55	0.97
Lower 40%	11.99	4.84	49.44	6.46	4.24	6.89	0.63	1.39	1.62	1.71	0.78
Upper 30%	17.98	8.57	58.87	10.55	4.54	7.64	1.72	2.82	3.59	3.96	1.37

Source: SUSENAS 1978 data tapes, Biro Pusat Statistik, Jakarta.

Table 15: SOURCES OF NUTRIENTS, OUTER ISLANDS, INDONESIA, 1978

(%)

Expenditure Group	Calories	Protein	Fat	Carbohydrates	Calcium	Iron	Vitamin A	Thiamine	Riboflavin	Niacin	Vitamin C
Rice - Total	60.80	46.40	11.80	70.80	11.32	32.51	0.00	52.11	29.67	53.95	0.00
Lower 40%	61.65	49.33	15.31	70.23	12.83	34.50	0.00	54.01	31.35	56.24	0.00
Upper 30%	57.84	41.48	8.26	69.28	9.14	28.89	0.00	47.78	26.33	49.60	0.00
Corn - Total	2.47	2.66	2.24	3.02	1.37	3.12	5.62	3.94	3.31	2.75	1.41
Lower 40%	3.85	4.27	3.78	4.70	2.52	4.77	8.24	5.69	5.33	4.56	2.38
Upper 30%	1.64	1.63	1.19	1.99	0.60	2.07	3.74	2.95	1.88	1.52	0.51
Wheat - Total	0.96	1.26	0.42	1.12	0.89	1.75	0.22	2.90	1.04	1.34	0.00
Lower 40%	0.53	0.66	0.34	0.61	0.51	0.91	0.29	1.44	0.59	0.66	0.00
Upper 30%	1.47	1.94	0.54	1.76	1.29	2.73	0.09	4.66	1.54	2.13	0.00
Cassava - Total	4.76	1.61	1.02	5.81	7.30	4.70	0.00	3.89	1.00	2.43	9.53
Lower 40%	6.35	2.25	1.57	7.54	9.23	6.28	0.00	4.96	1.31	3.30	10.27
Upper 30%	3.46	1.06	0.56	4.42	5.46	3.31	0.00	2.94	0.71	1.69	8.66
Potatoes - Total	2.02	1.42	0.96	2.39	3.32	2.77	8.77	3.49	2.56	2.15	4.66
Lower 40%	2.51	1.90	1.47	2.85	4.13	3.42	7.90	4.11	3.15	2.66	5.40
Upper 30%	1.66	1.03	0.54	2.10	2.70	2.27	10.28	3.04	2.08	1.76	4.19
Fish - Total	3.99	24.15	7.94	0.00	19.59	10.31	4.74	2.93	10.52	15.45	0.00
Lower 40%	3.42	21.69	7.56	0.00	18.24	9.09	4.79	2.37	9.67	13.25	0.00
Upper 30%	4.61	26.18	8.17	0.01	20.77	11.26	4.66	3.46	11.28	17.48	0.00
Meat & Poultry - Total	0.95	2.76	4.01	0.00	0.42	1.93	0.38	1.45	2.15	2.93	0.00
Lower 40%	0.45	1.37	1.98	0.00	0.28	0.92	0.25	0.65	1.05	1.40	0.00
Upper 30%	1.67	4.65	6.54	0.00	0.37	3.35	0.55	2.60	3.67	5.03	0.00
Eggs - Total	0.22	0.62	1.09	0.00	0.63	0.65	1.03	0.34	1.23	0.04	0.00
Lower 40%	0.15	0.43	0.90	0.00	0.48	0.44	0.88	0.23	0.85	0.02	0.00
Upper 30%	0.33	0.87	1.36	0.00	0.81	0.91	1.13	0.48	1.71	0.05	0.00
Dairy Products - Total	0.23	0.50	0.61	0.12	2.92	0.05	0.38	0.25	1.72	0.05	0.07
Lower 40%	0.07	0.17	0.26	0.03	0.93	0.02	0.13	0.08	0.54	0.02	0.02
Upper 30%	0.45	0.96	1.06	0.24	5.62	0.10	0.71	0.49	3.35	0.10	0.11
Vegetables - Total	3.46	7.48	2.34	4.17	38.25	25.31	66.27	16.92	37.82	11.36	71.32
Lower 40%	3.45	7.97	2.89	4.06	38.86	24.37	66.56	16.65	38.74	12.06	69.34
Upper 30%	3.51	7.05	1.84	4.36	36.49	25.97	64.04	17.05	36.33	10.70	72.12
Legumes - Total	1.87	5.35	6.29	0.95	6.70	4.93	0.83	7.05	3.50	3.83	0.84
Lower 40%	1.40	4.39	5.01	0.72	5.08	3.96	0.85	5.69	2.68	2.58	0.69
Upper 30%	2.51	6.69	7.91	1.27	8.69	6.24	0.92	8.89	4.62	5.44	1.13
Fruits - Total	1.67	0.77	0.64	2.30	2.68	3.24	10.12	2.36	3.27	1.69	9.49
Lower 40%	1.33	0.63	0.49	1.80	2.10	2.53	7.85	1.94	2.73	1.44	7.73
Upper 30%	2.13	0.96	0.82	2.97	3.35	4.12	12.44	2.93	3.92	2.04	11.51
Other - Total	16.56	4.98	60.61	9.28	4.57	8.70	0.86	2.35	2.15	1.97	1.71
Lower 40%	14.74	4.85	58.34	7.35	4.74	8.69	0.48	2.10	1.93	1.72	2.02
Upper 30%	18.72	5.49	61.22	11.60	4.44	8.79	1.29	2.73	2.51	2.40	1.62

Source: SUSENAS 1978 data tapes, Biro Pusat Statistik, Jakarta.

In addition to rice, the major sources of thiamine in Java are vegetables and legumes. There are no noticeable differences between the two major regions.

For riboflavin, vegetables again constitute a major source, next to rice. However, there is a noticeable difference between Java and the Outer Islands; in the latter, fish contributes relatively more to this nutrient, while legumes are more important in Java.

Niacin comes primarily from rice. In addition, it comes from vegetables and fish in the Outer Islands, and from legumes and cassava in Java.

Vitamin C is basically embodied in vegetables and fruits, as well as in cassava. Fruits are more important for the rich than for the poor as a source of vitamin C.

Given the importance of rice in the Indonesian diet, the intakes of all nutrients except for calcium, vitamin A and vitamin C are bound to be very sensitive to any changes in income and the price of rice, or any factor affecting them, even if its own income and price elasticities are low. This will be the case especially for the poor and in Java, where rice is a basic staple, accounting for a substantial portion of total household expenditures on food.

For similar reasons, but to a lesser extent, the prices of cassava and vegetables are particularly important in Java. Fish is significant in the Outer Islands, and so are legumes in Java as major sources of almost all other micronutrients.

4.2   Total Expenditure Elasticities of Demand for Foods and Nutrients

The total expenditure elasticities of demand for foods and nutrients are reported in Tables 16 and 17 respectively, with breakdowns by expenditure

Table 16: TOTAL EXPENDITURE ELASTICITIES OF DEMAND FOR FOOD, INDONESIA, 1978

Expenditure Group	Rice	Corn	Wheat	Cassava	Potatoes	Fish	Meat	Eggs & Poultry	Dairy	Vegetables Products	Legumes	Fruit	Other
Java													
Lower 40%	3.022 (16.410)	-0.622 (-2.52)	0.061 (0.672)	0.238 (0.912)	0.539 (4.398)	1.317 (8.120)	3.948 (5.561)	1.143 (7.453)	0.076 (2.377)	0.953 (10.649)	2.613 (14.257)	1.901 (10.998)	1.243 (21.753)
Middle 30%	0.914 (4.087)	-0.425 (-0.929)	-0.027 (0.148)	0.790 (1.454)	1.238 (3.884)	1.825 (5.374)	2.162 (8.196)	2.871 (5.935)	0.783 (3.900)	0.990 (95.369)	1.991 (5.248)	3.708 (8.031)	0.911 (9.573)
Upper 30%	0.034 (1.098)	0.203 (2.210)	0.943 (12.957)	-0.074 (-0.580)	1.673 (14.836)	0.979 (11.936)	2.534 (26.233)	2.544 (18.787)	2.203 (23.967)	0.559 (13.689)	0.653 (9.383)	2.617 (25.007)	0.696 (25.667)
Calorie-Deficient Population	0.903 (13.444)	-0.580 (-5.973)	0.194 (4.457)	-0.257 (-2.278)	0.964 (13.114)	0.823 (10.927)	1.680 (26.355)	2.416 (23.843)	1.274 (22.550)	0.615 (14.319)	1.370 (17.298)	2.768 (29.833)	0.798 (30.055)
Outer Islands													
Lower 40%	1.149 (14.565)	-0.290 (-2.041)	0.424 (5.243)	0.365 (2.002)	0.667 (4.833)	1.834 (16.703)	1.332 (13.692)	2.110 (13.037)	0.719 (10.229)	1.735 (19.505)	1.425 (10.348)	2.814 (16.958)	1.331 (28.731)
Middle 30%	0.398 (4.423)	-0.133 (-0.399)	1.224 (3.752)	0.717 (1.358)	1.666 (0.444)	1.638 (7.429)	3.112 (8.785)	3.621 (6.787)	1.856 (5.996)	0.895 (5.565)	2.362 (5.392)	4.288 (9.243)	0.945 (10.968)
Upper 30%	0.053 (2.330)	0.574 (5.453)	1.672 (13.391)	0.108 (0.694)	1.632 (11.190)	0.459 (8.650)	3.148 (24.570)	3.050 (18.672)	2.779 (24.140)	0.554 (15.510)	2.311 (18.044)	2.052 (17.976)	0.705 (32.250)
Calorie-Deficient Population	0.287 (6.162)	-0.223 (-2.373)	0.702 (8.851)	0.177 (1.348)	0.974 (8.680)	1.201 (17.843)	2.278 (25.109)	2.987 (22.922)	1.859 (22.999)	1.077 (19.763)	2.142 (19.811)	3.015 (25.375)	0.866 (27.840)

Note: t-Statistics are given in parentheses.

Source: Data tapes of SUSENAS 1978, Biro Pusat Statistik, Jakarta.

Table 17: TOTAL EXPENDITURE ELASTICITIES OF DEMAND FOR NUTRIENTS, INDONESIA, 1978

	Calories	Protein	Fat	Carbohydrates	Calcium	Iron	Vitamin A	Thiamine	Riboflavin	Niacin	Vitamin C
Java											
Lower 40%	0.789 (21.750)	0.914 (29.203)	1.224 (23.746)	0.702 (19.776)	0.805 (15.629)	0.759 (21.624)	0.992 (5.768)	0.933 (27.327)	0.753 (23.622)	0.790 (29.497)	0.876 (6.217)
Middle 30%	0.543 (7.771)	0.682 (10.041)	0.952 (9.155)	0.479 (7.116)	0.900 (8.878)	0.660 (9.678)	1.535 (4.844)	0.652 (9.528)	0.642 (10.561)	0.559 (9.513)	1.450 (5.285)
Upper 30%	0.298 (12.541)	0.424 (19.388)	0.604 (22.526)	0.218 (9.973)	0.611 (22.843)	0.438 (21.287)	0.836 (16.148)	0.366 (20.092)	0.507 (28.084)	0.362 (18.561)	0.820 (16.753)
Calorie-Deficient Population	0.304 (14.555)	0.410 (21.057)	0.678 (26.125)	0.237 (12.246)	0.538 (21.019)	0.386 (21.724)	0.477 (13.645)	0.411 (24.636)	0.436 (27.371)	0.315 (19.395)	0.834 (13.582)
Outer Islands											
Lower 40%	0.617 (18.483)	0.788 (26.622)	1.122 (25.999)	0.556 (17.622)	0.846 (22.147)	0.729 (25.983)	1.736 (13.558)	0.695 (27.311)	0.762 (31.139)	0.694 (27.113)	1.434 (12.497)
Middle 30%	0.535 (15.398)	0.666 (18.348)	0.920 (12.466)	0.472 (12.013)	0.847 (13.479)	0.726 (16.443)	1.199 (7.05)	0.640 (14.222)	0.691 (15.727)	0.621 (17.067)	0.561 (3.119)
Upper 30%	0.288 (16.474)	0.436 (26.291)	0.597 (27.845)	0.213 (12.477)	0.626 (28.402)	0.457 (28.038)	0.632 (15.347)	0.429 (27.464)	0.569 (30.812)	0.421 (23.696)	0.512 (13.645)
Calorie-Deficient Population	0.212 (8.867)	0.370 (18.106)	0.702 (24.593)	0.144 (6.482)	0.607 (22.456)	0.411 (21.405)	1.116 (14.910)	0.336 (20.632)	0.472 (27.755)	0.308 (16.461)	0.835 (12.637)

Note: t-Statistics are given in parentheses.

Source: Data tapes of SUSENAS 1978, Biro Pusat Statistik, Jakarta.

class; separate parameter estimates are given for the population deficient in calories.

Among the staples, including potatoes, it is clear that rice, the major staple, has the highest income elasticity of demand among low-income groups in both Java and the Outer Islands. This is particularly true for Java, where the lower 40% expenditure group is poorer than in the Outer Islands. On the other hand, among the higher-expenditure groups we find higher expenditure elasticities for corn, wheat, and potatoes. This under-lines the fact that these staples are consumed by the higher-income groups. Cassava consumption appears not to respond to any changes in income in both regions. The calorie-deficient population in Java appears, with regard to consumption of rice and corn, to have the attributes of the middle income group. It has a relatively high price elasticity for rice and a low one for cassava.

These estimates show that, as household total expenditure increases, people at low incomes increase rice consumption and eventually switch to some corn, wheat and potatoes. The calorie-deficient population is inclined to increase their consumption of rice and wheat with rising incomes, and not the consumption of corn and cassava in Java. Cassava appears more appealing to the calorie-deficient population in the Outer Islands.

Among the major sources of protein, namely fish, meat, poultry, milk, eggs, and legumes we find relatively high income elasticities of demand for fish and legumes among the lower income groups, and for meat, poultry, dairy products and eggs, the more expensive items, among the higher-income groups. This marks a relative shift from fish and vegetables to the other sources of protein as income rises.

Among the major sources of micronutrients, namely legumes, vegetables fruits and "other" foods, we observe almost uniformly high income elasticities of demand for fruits, especially among the middle classes, and for "other" foods and legumes among the lower classes.

The total expenditure elasticities of demand for nutrients are not easy to predict from the previous discussion, as these combine the effects of total expenditures on all foods, as well as the relative contributions of various foods to different nutrients. The findings are reported in Table 17.

What is commonly observed are the falling elasticities by income groups. This signifies saturation in terms of nutrients, and underscores the previous findings that households shift to more expensive foods as their income rises, while maintaining their nutrient consumption. That is, while changes in total expenditures bring about some dramatic changes in food consumption, the changes in nutrient consumption are much less pronounced.

Among low-income groups, fat stands out with a high income elasticity of demand. Likewise vitamin A and vitamin C in the Outer Islands. The finding concerning fats may signify the substitution, which is apparent among the poor, from fish to meat as income rises, as well as the dependence of the poor on legumes. Vitamin A and vitamin C are known to be income-sensitive micronutrients, since they are so closely related to the consumption of particular fruits and vegetables.

The relatively lower elasticities, especially for the lower-income group, in the Outer Islands, reflects the fact that the lower 40% expenditure group there is better-off than the lower 40% expenditure group in Java. In both Java and the Outer Islands, changes in incomes of the calorie-deficient population are most likely to affect their consumption of fats, vitamins A and C, and calcium.

4.3   Price Elasticities of Demand for Foods and Nutrients

We turn now to a discussion of own and cross (compensated) price elasticities of demand for foods and nutrients in which we focus on the population deficient in calories who are of the most concern to policy makers. Table 18 gives price elasticities of demand for different foods for the calorie-deficient groups for Java and the Outer Islands separately. [1]

All of the own price elasticities of demand, except for legumes and corn in Java, are negative, as would be expected. In Java, the consumption of potatoes, fish, meat and poultry, dairy products and fruits is especially sensitive to changes in their own prices; this is the case for corn, wheat, potatoes, meat and poultry, dairy products, eggs, legumes and fruits in the Outer Islands. Relatively inelastic demand for corn, vegetables and "other" food items with respect to their own prices is observed in Java, and for rice and "other" food items in the Outer Islands.

The price elasticity of demand for rice is higher in Java than in the Outer Islands; a 10% increase in the price of rice will bring about a 15% decline in the quantities consumed by the calorie-deficient households in Java, but only an 8% decline in the Outer Islands, where the population is less dependent on rice.

In interpreting the effects of price changes on other commodities, we focus on those cases where the cross price elasticities are not only statistically significant but also relatively large. First, among the staples we clearly observe some substitution among them, although the cross price elasticities are not always consistent in magnitude and sign. An increase in the price of rice leads to greater consumption of corn in Java. An increase in the price of corn has a particularly strong positive impact on consumption of

_____

[1]   Only statistically significant coefficients are presented.

Table 18: PRICE ELASTICITIES OF DEMAND FOR FOODS, FOR POPULATION DEFICIENT IN CALORIES, INDONESIA, 1978

Region/ Food Category	Rice	Corn	Wheat	Cassava	Potatoes	Fish	Meat and Poultry	Eggs	Dairy Products	Vegetables	Legumes	Fruits	Other
Java													
Own Price Elasticity	-1.48	0.52	-0.74	-1.36	-3.35	-1.95	-1.86	-1.83	-2.75	-0.81	1.07	-2.40	-0.55
Cross Price Elasticities:													
Rice	*	2.69	1.13	1.43	-0.78	-1.45	1.66	0.91	0.72	-1.07	0.73		
Corn	0.38	*	0.32	0.73	0.56	3.33	0.63	0.66		0.31		-1.27	-0.29
Wheat	2.49	-4.36	*		3.00		-0.39		-0.43	0.12		1.49	
Cassava	-0.79	1.33	0.61	*	0.82	-0.30			0.33				
Potatoes					*				-.48				
Fish	-0.22	-0.11	-0.25	-0.50			-.44		0.16				-0.15
Meat and Poultry	-0.12	-1.34	-0.53	-1.72	-.25	-0.89	*	-.40	-.76		-1.57		
Eggs				-.61	1.15	2.16		*	-0.85			-0.68	-0.26
Dairy Products	0.16							-0.51	8			-0.63	
Vegetables	0.27	-0.66						0.27		*	-.13	0.16	
Legumes		-0.46	0.16	0.28	0.19		0.14	0.28		-.13	*	0.29	0.65
Fruits				-0.29			0.30		0.16			*	-0.08
Other		-0.35	-0.06	-0.55	-0.13	-0.21	-0.27	-0.24		-0.49	-0.15		*
Outer Islands													
Own Price Elasticity	-0.83	-3.28	-1.82	-1.47	-2.82	-1.20	-2.31	-2.88	-2.36	-1.01	-2.24	-2.14	-0.60
Cross Price Elasticities:													
Rice	*	1.12		1.05	0.65			-1.12					
Corn	0.78	*		2.16	1.11	-0.55	-0.89	-0.74		-0.48	-0.58	-1.15	-0.23
Wheat	-0.39	1.39	*	1.03	2.62	0.35		1.73	0.89		-.87	1.38	0.25
Cassava	-0.30		0.31	*	0.85		-0.41	-0.46		-0.27	-0.51	0.36	-0.11
Potatoes					*	0.21							
Fish	0.24	-0.39		-0.32	0.67	*	0.68	0.55		0.22	0.14		
Meat and Poultry		-1.36		-0.51	1.64	0.81	*	-0.52			0.40		-.10
Eggs				0.65	-0.74			*	0.72	0.61			
Dairy Products	-0.85	-0.61		-1.85	1.14	-0.72			*	0.17		-0.89	
Vegetables	-0.12	0.25		-0.30		0.26				*	-0.27	-0.85	
Legumes		-0.48	-.87			0.14	0.40			-0.12	*		
Fruits				-0.63	0.26		-0.23	-0.37	-0.35			*	-0.59
Other	0.20	-0.27		-0.38	-0.25	-0.19	0.25	0.22		-0.20	-.63		*

Note: Statistically insignificant coefficients have been omitted.

Source: Data tapes of SUSENAS 1978, Biro Pusat Statistik, Jakarta.

cassava and potatoes, especially in the Outer Islands. An increase in the price of wheat leads to increasing consumption of cassava and potatoes, as well as rice in the Outer Islands. Hence, there is a fair amount of substitution among the staples.

Second, what is perhaps more interesting is the extent to which changes in the prices of staples affect the consumption of nonstaple foods, and vice versa, as well as the way in which the consumption of different nonstaple foods is affected by the prices of other nonstaples. The food groups the consumption of which is most sensitive to price changes in both the prices of staples and of other nonstaples are fish, fruits and vegetables in Java; and fruits in the Outer Islands. Unlike the case of fish, meat and poultry consumption is not as sensitive to the prices of any of the other food groups. The nature of the effects relates to specific food consumption patterns which are hard to study with these data. It is not clear, for example, why consumption of meat and poultry in Java is positively correlated with the price of rice. One effect may be related to the correlation between consumption of expensive varieties of rice by the rich and consumption of meat and poultry.

In the Outer Islands, the situation is quite different in that the cross price elasticities tend to be lower than in Java, so that the consumption of a food group is less sensitive to changes in the prices of the other foods. This may reflect the fact that the population in the Outer Islands is better-off than that in Java and therefore food prices matter less in the latter region.

The consumption of nutrients is affected by price changes through their impact on food consumption, given the shares in the total food budget of the food groups which are significantly affected. The price elasticities of demand for nutrients are presented in Table 19. As would be expected, given

Table 19: PRICE ELASTICITIES OF DEMAND FOR NUTRIENTS, FOR POPULATION DEFICIENT IN CALORIES, INDONESIA, 1978

Region/Food Category	Calories	Protein	Fat	Carbohydrates	Calcium	Iron	Vitamin A	Thiamine	Riboflavin	Niacin	Vitamin C
Java											
Rice	-0.35	-0.27		-0.38	0.25	0	-.90	-0.13		-0.41	0.51
Corn	0.08	0.14	0	0	0	0	0	0	0	0.16	
Wheat	0.31	0.18	-0.29	0.33	-0.27	0.11	-1.03		-0.25	-0.21	-1.01
Cassava	-0.09		-.16	-0.10	0.11	0.81	0.76		-.1-	-0.09	
Potatoes	-0.09			-0.11		-0.06	-0.73	-0.07		-0.07	-0.17
Fish		-0.14						-0.06		-0.09	
Meat & Poultry		-0.13	-0.26							-0.02	
Eggs						-0.18					
Dairy Products										0.20	
Vegetables		-0.04	0		-0.18	-0.12	-0.36	-0.28	-0.01	-0.42	
Legumes		-0.06		0	-0.07	-0.03	0.15	-0.04	-0.04		0.15
Fruits						-0.06					-0.20
Other	-0.08	-0.11	-0.22	-0.07	-0.22	-0.15	-0.66	-0.08	-0.13	-0.08	-0.60
Outer Islands											
Rice	-0.20	-0.17	-0.02	-0.24	0.21			-0.14		-0.22	0.56
Corn				0.11							
Wheat											
Cassava	-0.10			-0.12	-0.21	-0.07	-0.28	-0.07	-0.04	-0.05	-0.43
Potatoes				-0.04	-0.07	-0.06	-0.36	-0.08	-0.03	-0.03	-0.24
Fish	-0.04	-0.12			-0.06			-0.05			
Meat & Poultry			0.16	0.07	0.20	-0.09	0.79	-0.07	0.16		0.86
Eggs			-0.09			-0.06					-0.28
Dairy Products									0.07		-0.29
Vegetables		-0.02	0	-0.03	-0.20	-0.12	-0.31	-0.10	-0.23	-0.05	-0.19
Legumes		-0.09	-0.13		-0.23	-0.13	-0.17	-0.17	-0.10		-0.18
Fruits			-0.07			-0.04	-0.17			-0.04	-0.21
Other		-0.05	-0.04		-0.13	-0.10	-0.19	-0.05	-0.08	-0.04	-0.33

Note: Statistically insignificant coefficients have been omitted.
Source: Data tapes of SUSENAS 1978, Biro Pusat Statistik, Jakarta.

the importance of rice as a source of nutrients, a rise in its price adversely affects the intakes of calories, protein, carbohydrates, thiamine and niacin, especially in Java. However, through the substitution of other foods, there is in fact a net increase in the consumption of vitamin A and calcium in Java, and vitamin C in both regions.

The prices whose rises bring about negative effects on the consumption of nutrients are the price of wheat, on the consumption of calcium in Java; the price of cassava, on the consumption of calcium in the Outer Islands; the price of vegetables, on the consumption of iron and riboflavin; the price of cassava and potatoes, on the consumption of vitamin A in the Outer Islands; and the prices of cassava and wheat, on the consumption of vitamin C.

4.4   Size Elasticities of Demand for Foods and Nutrients

Next we turn to the effect of the household's demographic character-istics on food consumption and nutrition. These effects are expected to be similar to those from changes in income which constrain the household's budget. [1] The findings for the calorie-deficient population of Java and the Outer Islands are presented in Tables 20 and 21.

The results exhibit some general and expected patterns. As the number of household members increases, households reduce their consumption of meat and poultry, eggs, and fruits, as well as legumes in the Outer Islands. They do so in order to increase consumption of all other categories of food; there is a notable relative increase in consumption of cassava.

---

[1]   Unlike the previous estimates, these are not elasticities, but coeffi-cients of a semi-logarithmic relationship; the marginal effect of an increase in the independent variable rises with the rise in the dependent variable.

Table 20: ESTIMATED EFFECTS OF FAMILY SIZE AND COMPOSITION ON FOOD CONSUMPTION, POPULATION DEFICIENT IN CALORIES, INDONESIA, 1978

Region/Age Group	Rice	Corn	Wheat	Cassava	Potatoes	Fish	Meat & Poultry	Eggs	Dairy Products	Vegetables	Legumes	Fruits	Other
Java													
Ages 0-4	0.087 (2.403)	0.097 (1.843)	-0.005 (-0.24)	0.275 (4.49)	-0.046 (-1.17)	0.159 (3.909)	-0.145 (-4.198)	-0.158 (12.881)	0.274 (8.951)	0.170 (7.302)	0.027 (0.634)	0.038 (0.756)	0.064 (4.455)
Ages 5-9	0.098 (2.25)	0.115 (1.822)	0.029 (1.031)	0.356 (4.851)	-0.032 (-0.689)	0.091 (1.873)	-0.064 (-1.553)	-0.135 (-2.055)	-0.186 (-5.059)	0.103 (3.704)	0.014 (0.285)	-0.219 (-3.63)	0.023 (1.365)
Ages 10+	0.041 (1.749)	0.205 (5.959)	0.009 (0.584)	0.311 (7.753)	0.017 (0.67)	0.049 (1.858)	-0.112 (-4.461)	-0.196 (-5.468)	-0.154 (-7.697)	0.150 (9.872)	0.034 (1.21)	-0.338 (-10.282)	0.037 (3.926)
Outer Islands													
Ages 0-4	0.098 (5.028)	0.105 (2.687)	0.056 (1.687)	0.199 (3.639)	0.081 (1.744)	0.025 (0.902)	-0.169 (-4.459)	-0.23 (-4.227)	0.072 (2.155)	0.058 (2.584)	-0.153 (-3.394)	0.027 (0.553)	0.055 (4.242)
Ages 5-9	0.174 (7.795)	-0.046 (-1.039)	0.02 (0.53)	0.009 (0.143)	0.013 (0.259)	0.055 (1.711)	-0.114 (-2.62)	-0.39 (-6.241)	-0.146 (-3.79)	0.018 (0.716)	-0.11 (-2.135)	-0.183 (-3.211)	0.054 (3.626)
Ages 10+	0.150 (12.12)	0.111 (4.464)	-0.009 (-0.459)	0.114 (3.272)	0.08 (2.676)	-0.007 (-0.409)	-0.127 (-5.263)	-0.254 (-7.327)	-0.095 (-4.447)	0.041 (2.879)	-0.087 (-3.024)	0.055 (4.242)	0.040 (4.928)

Note: t-statistics are given in parentheses.

Source: Data tapes of SUSENAS 1978, Biro Pusat Statistik, Jakarta.

Table 21: ESTIMATED EFFECTS OF FAMILY SIZE AND COMPOSITION ON NUTRIENT CONSUMPTION, POPULATION DEFICIENT IN CALORIES, INDONESIA, 1978

Region/Age Group	Calories	Protein	Fat	Carbohydrates	Calcium	Iron	Vitamin A	Thiamine	Riboflavin	Niacin	Vitamin C
Java											
Ages 0-4	0.138 (12.174)	0.130 (12.352)	0.11 (7.897)	0.144 (13.76)	0.141 (10.167)	0.116 (12.065)	0.202 (5.202)	0.116 (12.918)	0.127 (14.741)	0.134 (15.201)	0.195 (5.861)
Ages 5-9	0.148 (10.879)	0.124 (9.827)	0.091 (5.401)	0.159 (12.704)	0.091 (5.464)	0.126 (10.918)	0.079 (1.702)	0.122 (11.293)	0.09 (8.684)	0.134 (12.670)	0.094 (2.363)
Ages 10 +	0.167 (22.478)	0.148 (21.473)	0.099 (10.758)	0.180 (26.277)	0.128 (14.107)	0.149 (23.734)	0.157 (6.194)	0.147 (24.861)	0.126 (22.399)	0.161 (27.919)	0.163 (7.473)
Outer Islands											
Ages 0-4	0.114 (11.406)	0.088 (10.316)	0.076 (6.444)	0.119 (12.817)	0.094 (8.404)	0.092 (11.524)	0.115 (3.677)	0.099 (14.572)	0.086 (12.177)	0.094 (12.026)	0.155 (5.624)
Ages 5-9	0.146 (12.734)	0.120 (12.248)	0.063 (4.614)	0.159 (14.91)	0.062 (4.826)	0.103 (11.216)	0.04 (1.134)	0.122 (15.605)	0.079 (9.726)	0.130 (14.519)	0.037 (1.169)
Ages 10 +	0.151 (23.782)	0.120 (22.028)	0.074 (9.815)	0.163 (27.487)	0.085 (11.961)	0.111 (21.887)	0.071 (3.564)	0.128 (29.579)	0.094 (20.809)	0.130 (26.098)	0.096 (5.501)

Note: t-statistics are given in parentheses.

Source: Data tapes of SUSENAS 1978, Biro Pusat Statistik, Jakarta.

As expected, with the advent of children, households consume more dairy products; these are given up when the household has a relatively larger number of adults who consume more corn and cassava. These shifts in food consumption due to variations in family size and composition do not bring about fundamental changes in the consumption of nutrients, as shown in Table 21.

Cassava has a relatively important role in protecting the consumption of nutrients of larger families, although not enough to avoid a decline in terms of nutrients per capita. Nutrients which are particularly vulnerable are protein and fat, iron for children, and vitamins A and C for adults. On the whole, the Outer Islands are more vulnerable to inadequate nutrient consumption as a result of increasing family size.

4.5   Efficiency in Food and Nutrition Consumption

Some of the major practical and conceptual issues related to food consumption concern food quality and efficiency of nutrient consumption. As is evident in the data, wealthier households buy more expensive food varieties. In fact, the estimated price elasticities incorporate a degree of quantity-quality substitution, whereby households consuming more expensive foods buy smaller quantities than households consuming relatively inexpensive foods.

There are many explanations to this behavior in terms of taste, such as variety or packaging, and possible time saving in preparation. As pointed out earlier, however, a more expensive diet does not necessarily imply a better diet in terms of nutrition. [1] Hence, from a narrow perspective of the relation between food consumption and nutrition, one can pose the question of how efficiently households attain their nutritional requirements. In other words, for given incomes and market prices, are some households better than others in attaining certain levels of nutrient intakes?

We first address this question by looking at the prices that people pay for a given amount of a particular nutrient embodied in an average basket of foods comprising their diet. This is a function of food prices and the shares of foods in the provision of particular nutrients.

These average prices are presented in Table 22. The results show relatively expensive diets in Jakarta and relatively inexpensive diets in East Java. Diets are more expensive in urban areas compared with rural areas, and in November compared with the other months observed.

While the rich pay more for their diets than the poor, there is no evidence that the calorie-deficient population pays more than the nondeficient population. On the whole, protein and vitamins are derived from more expensive sources for the nondeficent population, compared with the calorie-deficient group.

This conclusion is supported by a set of unreported regressions in which the prices of nutrients have been used as dependent variables to be explained by household monthly expenditure, family size and composition, education of the spouse of the household head, and source of income. [2] Higher levels of expenditures and smaller numbers of household members, particularly

---

[1] Given the estimated level of calorie deficiency in Indonesia, we do not deal with the qualitative aspects of expensive diets, traces of which can be seen in the data which show higher consumption of fats and "other" foods, including soft drinks, with rising incomes.

[2] These regressions can be obtained upon request.

Table 22: AVERAGE "PRICES" OF NUTRIENTS EMBODIED IN A KILOGRAM OF FOOD

	Calories	Protein	Fat	Carbohydrates	Calcium	Iron	Vitamin A	Thiamine	Riboflavin	Niacin	Vitamin C
Indonesia	125	176	105	120	171	157	179	141	177	168	170
Region											
DKI Jakarta	170	261	166	158	279	255	283	211	270	249	290
West Java	133	184	110	128	175	156	154	138	167	169	140
Central Java	105	131	77	106	128	122	154	118	144	133	149
DI Yogyakarta	104	131	80	106	127	126	198	127	153	138	195
East Java	106	149	88	103	135	130	151	121	153	145	146
Sumatra	143	214	120	138	203	187	206	167	209	201	191
Bali & Nusatenggara	117	167	108	112	176	155	174	140	184	161	193
Kalimantan	151	239	155	138	251	222	277	184	247	223	238
Sulawesi	132	196	122	125	199	174	218	153	197	180	200
Maluku & Irian Jaya	146	227	133	138	227	217	318	191	257	213	258
Location											
Urban	155	225	146	146	222	215	239	184	228	216	232
Rural	118	166	97	115	160	145	167	133	167	158	157
Season											
February	125	169	100	122	166	152	167	139	168	163	160
May	117	167	102	113	170	153	180	136	172	158	169
November	132	192	114	126	171	166	192	150	193	184	180
Expenditure Class											
Lower 40%	109	148	84	108	147	130	155	123	151	140	147
Middle 30%	124	171	100	121	169	152	175	139	173	163	163
Upper 30%	145	216	137	135	202	196	214	168	215	208	204
Nutritional Status											
Calories: Deficient	121	165		118	164	149	173	136	167	157	165
Not Deficient	129	110	112	123	179	166	187	148	189	181	174
Protein: Deficient	113	153	89	112	52	131	167	129	159	145	160
Not Deficient	132	191	116	120	183	170	187	150	189	183	175
Vitamin A: Deficient	123	169	102	120	169	151	190	137	171	160	105
Not Deficient	126	182	108	121	172	162	171	145	183	175	173

Source: SUSENAS 1978 data tapes, Biro Pusat Statistik, Jakarta

of adults, are associated with higher prices for nutrients. The prices of iron and vitamin A appear to be particularly sensitive to variations in income and family size.

Education has a somewhat similar effect to the level of expenditures. That is, higher levels of education are in general associated with more expensive diets in Java, and to a lesser degree in the Outer Islands.

As far as sources of income are concerned, households reporting incomes from agriculture also report the least-cost diets, all other things equal. This presumably reflects the fact that people engaged in agriculture have easier access to food.

These results suggest that the poor can to a degree "protect" their nutritional intake by obtaining their food from relatively inexpensive sources. But to the extent that the findings of previous sections indicate clearly that lower food prices in themselves do not give enough "protection" for the poor, it is important to study the effect of education and sources of income on the diet while incomes, prices, and family size and composition are controlled for.

The results reported in Tables 23 and 24 show that in both Java and the Outer Islands higher education is associated with a reduction in the consumption of calories, protein, carbohydrates, iron, thiamine and niacin. Interestingly enough, there does not seem to be a compensating gain in any other particular nutrient. This finding, combined with that concerning prices, strongly suggests that the more educated homemakers buy more expensive foods, but at a net loss in terms of nutrition. The exception is the higher consumption levels of fats among families of homemakers with elementary schooling.

Table 23: REGRESSION COEFFICIENTS ON EDUCATION OF SPOUSE OF HEAD OF HOUSEHOLD, WITH CONSUMPTION OF NUTRIENTS AS DEPENDENT VARIABLES

Region/Level of Education	Calories	Protein	Fat	Carbohydrates	Calcium	Iron	Vitamin A	Thiamine	Riboflavin	Niacin	Vitamin C
Java											
Education of Spouse of Head:											
Elementary School	*	*	0.1245 (3.3858)	*	*	*	*	*	*	*	*
Junior High School	-0.1397 (-2.1301)	*	*	-0.1409 (-2.2220)	*	*	*	-0.1264 (-2.1895)	*	-0.1303 (-2.3581)	*
Senior High School	-0.1939 (-2.5828)	-0.1654 (-2.1822)	*	-0.1755 (-2.0203)	*	-0.1800 (-2.6201)	*	-0.2570 (-3.8901)	*	-0.1802 (-2.8475)	*
Higher Education	*	*	*	*	*	-0.3482 (-2.0535)	*	-0.3920 (-2.4103)	*	*	*
Outer Islands											
Education of Spouse of Head:											
Elementary School	*	*	0.0699 (3.1020)	*	*	*	*	*	*	*	*
Junior High School	-0.1011 (-2.7460)	*	*	-0.0924 (-2.5404)	*	-0.0727 (-2.2474)	*	-0.0982 (-3.0926)	*	-0.0956 (-3.1083)	*
Senior High School	-0.1445 (-3.0872)	-0.1022 (-2.5126)	*	-0.1524 (-3.2755)	*	-0.1028 (-2.5002)	*	-0.1078 (-2.6716)	*	0.1199	*
Higher Education	-0.2358 (-2.0350)	*	*	-0.2893 (-2.4806)	*	-0.2113 (-2.0673)	*	-0.2610 (-2.6111)	*	-0.2022 (-2.0905)	*

SOURCE: SUSENAS 1978 data tapes, Biro Pusat Statistik, Jakarta.
NOTE: * Statistically insignificant coefficient; income, prices of foods, sources of income and family size are controlled for; t-statistics in parentheses.

Table 24: REGRESSION COEFFICIENTS ON SOURCES OF INCOME, WITH CONSUMPTION OF NUTRIENTS AS DEPENDENT VARIABLES

	Calories	Protein	Fat	Carbohydrates	Calcium	Iron	Vitamin A	Thiamine	Riboflavin	Niacin	Vitamin C
Java											
Source of Household Income:											
Agriculture	0.1296 (3.5184)	0.1103 (2.9619)	*	0.1445 (4.0209)	*	0.0739 (2.1848)	*	*	0.1099 (3.5622)	0.1669 (5.3744)	*
Industry	0.1037 (2.9050)	0.0783 (2.1705)	*	0.1178 (3.3851)	0.1051 (2.3373)	0.0901 (2.7491)	*	0.0918 (2.9202)	0.0740 (2.4755)	0.0792 (2.6328)	0.2297 (2.0912)
Government	*	*	*	*	*	*	*	*	*	*	*
Other	-0.1086 (-2.0438)	-0.0764 (-1.4232)	*	*	*	-0.0976 (-2.0012)	*	*	*	*	*
Outer Islands											
Source of Household Income:											
Agriculture	0.1747 (9.4535)	0.0932 (5.7950)	0.0560 (2.3233)	0.2076 (11.1598)	0.0837 (3.7416)	0.1279 (7.8441)	*	0.1626 (10.2015)	0.1085 (7.1841)	0.1714 (11.1072)	0.2612 (4.5557)
Industry	0.0592 (2.0402)	*	*	0.0789 (2.7038)	*	*	*	*	0.0509 (2.1488)	0.0567 (2.3427)	0.1817 (2.0201)
Government	*	*	*	*	*	*	*	*	*	*	*
Other	*	*	*	*	*	*	-0.1551 (-2.0960)	*	*	*	*

SOURCE: SUSENAS 1978 data tapes, Biro Pusat Statistik, Jakarta.

NOTE: * Statistically insignificant coefficient; income, prices of foods and family size are controlled for; t-statistics in parenthesis.

In conclusion, households engaged in agricultural and industrial activities have better diets in terms of nutrition. The relatively educated homemakers have more expensive diets at some sacrifice of a broad range of nutrients. This last point strongly suggests that there is scope for nutrition education in Indonesia, especially in Java where nutritional inadequacy is less related to income than in the Outer Islands.

V.   SUMMARY AND CONCLUSION

This paper has two objectives. The first is to show how food and nutrition consumption in Indonesia vary by region, location of residence, season, and socioeconomic class. The second is to identify and estimate the basic parameters which explain why households consume what they do in terms of foods and nutrients, given their characteristics and market conditions. Particular attention has been given to the estimation of income and price elasticities of demand; these are good indicators of how households will respond to changes in incomes and prices, given their demographic and other characteristics. Understanding these responses is critical for the design of policies related to nutrition.

The study is based on the 1978 National Socio-Economic Survey (SUSENAS) which was conducted in four rounds and covered some 6,300 households in each round. Among others, information concerning the quantities of, and expenditures on, roughly 120 food items was collected.

We observe in the data great variations in food consumption patterns. The share of rice in household budgets ranges from 17% in Maluku and Irian Jaya to 44% in West Java. Cassava constitutes a relatively minor item in the household budget. The vast majority of the population report consumption of rice, vegetables and "other" foods. Fish, fruits, and legumes are relatively important as well.

Urban dwellers and the rich tend to have more varied diets. In this context it is important to point out that, while the budget share on staples declines with rising incomes, the share spent on rice increases, while that on corn, for example, decreases. Spatial and seasonal variations in food consumption patterns are highly consistent with observed variations in income levels and prices.

Consumption of different nutrients exhibits similar patterns to food consumption. Consumption of nutrients is generally higher in the relatively more affluent Outer Islands than in Java. Generally high consumption levels are reported in Sumatra and Kalimantan, and low levels in Central Java and Yogyakarta. Calories, carbohydrates and vitamin C are consumed in larger quantities in rural areas than in urban areas. Protein and vitamin A consumption is higher in urban Java than in rural Java. The opposite holds for the Outer Islands.

Although data on reported consumption should be interpreted with caution, the data strongly suggest widespread deficiencies in all nutrients. The problem is more serious in Java than in the Outer Islands, at least as far as calories are concerned. Furthermore, it appears that the problem is more one of maldistribution than of an overall shortfall in the availability of foods.

The major distinguishing characteristic between households which meet minimal nutritional requirements and those which do not is the level of household expenditure, especially in the Outer Islands. Households not deficient in calories tend to be smaller and have more educated homemakers.

In examining the sources of the various nutrients, the significance of rice as a contributor of most nutrients is striking. No other food group is so important from a nutritional standpoint in the Indonesian diet.

Variations in household incomes/expenditures are associated with marked variations in food consumption patterns. Except for the consumption of corn and cassava, households increase their consumption of all foods as they become wealthier. This is particularly true for wheat, potatoes, meat and poultry, milk, legumes and fruits. The same is true for nutrients, although the changes are much less dramatic than for foods.

These particular findings, combined with the effects on consumption patterns of household composition, indicate that the substitutions taking place in food consumption with changes in income and family size and composition are fairly efficient in "protecting" the nutritional intake of households.

The results concerning the effects of prices on food consumption patterns are relatively tentative, as these prices reflect in part qualitative differences in food consumption which require further study. Nonetheless, it is already apparent that there is a great deal of substitution going on as prices change. Among staples, consumption of rice among low-income groups in particular is relatively sensitive to changes in its own price; people tend to substitute rice with corn and cassava, and to a relatively lesser degree with wheat. Among the relatively poor, consumption of potatoes, fish, meat and fruits is relatively sensitive to changes in the prices of these commodities.

The substitution taking place when prices change is also manifested in the effect of price variation on the consumption of nutrients. We again see changes which are relatively smaller than those taking place in food consumption as prices vary. Consumption of calories, protein, and carbohydrates is adversely affected by an increase in the price of rice to such an extent that it can hardly be compensated for by lowering the price of any particular food group.

Relatively wealthy households with educated homemakers buy relatively expensive diets with some sacrifice in terms of nutrients.

Households which report incomes from agriculture are better-off, in terms of nutritional intake, than households with incomes from other sources. This suggests that agricultural households, which are relatively poor, also have easier access to their food.

To conclude, the results suggest that there is wide scope for nutrition policies based on changes in incomes and relative prices, as food and nutrient consumption respond rather dramatically to such changes. These may come about both in terms of the numbers of households consuming different food items and the quantities consumed of various foods. It is not clear at this stage how people respond to income and price changes in terms of buying more or less expensive food varieties, and what effect this has on nutrition.

As incomes rise, the dependency on rice increases as well. This dependency can be reduced by appropriate pricing policies. However, any such policies must take into account the evidence suggesting that no single food can substitute for rice as a major source of most nutrients. For example, inducing more cassava consumption at the expense of rice may increase calorie consumption but could at the same time be detrimental in terms of the intakes of other nutrients.

It is clear that the problem of nutritional adequacy in Indonesia is closely associated with levels of income and rice consumption. However, the data strongly suggest that inadequate diets are prevalent among the better-off and the better-educated as well; some of them consume more expensive diets at a loss in terms of nutrition. Hence, alleviating malnutrition in Indonesia is not just a matter of raising levels of income but also of nutrition education.

Appendix Table

Table A.1: PER CAPITA DAILY CONSUMPTION OF FOODS, FOR HOUSEHOLDS REPORTING CONSUMPTION, BY REGION AND LOCATION, INDONESIA, 1978
(in grams except for eggs which are numbers)

Region/Location	Rice	Corn	Wheat	Cassava	Potatoes	Fish	Meat & Poultry	Eggs	Dairy Products	Vegetables	Legumes	Fruits	Other
DKI Jakarta	300	42	19	47	32	29	26	0.236	22	95	62	86	241
Urban	300	42	19	47	32	29	26	0.236	22	95	62	86	241
West Java	402	86	23	129	69	29	31	0.184	20	136	47	93	234
Urban	345	50	28	75	42	28	31	0.242	26	130	68	121	279
Rural	409	89	21	132	74	29	31	0.167	17	137	44	89	229
Central Java	270	267	147	209	130	15	20	0.166	19	151	47	74	202
Urban	263	67	18	73	53	16	19	0.225	26	129	72	97	208
Rural	270	275	172	219	145	15	21	0.146	14	154	43	69	202
DI Yogyakarta	209	224	13	207	74	8	14	0.169	15	145	38	53	224
Urban	247	173	16	67	37	12	16	0.224	22	138	59	73	196
Rural	202	225	10	217	85	7	13	0.150	10	146	35	47	229
East Java	241	234	101	206	97	22	21	0.148	21	156	50	85	177
Urban	264	112	35	75	43	23	22	0.164	24	140	65	75	197
Rural	237	239	110	222	113	22	21	0.142	18	158	47	88	174
Sumatra	406	108	52	179	92	68	36	0.181	18	150	52	112	247
Urban	349	49	39	90	47	68	28	0.191	19	148	52	98	209
Rural	418	114	57	194	104	68	38	0.177	18	151	52	116	256
Bali & Nusatenggara	349	276	83	213	281	43	58	0.164	24	169	44	122	202
Urban	369	151	52	100	54	45	33	0.211	21	157	39	68	192
Rural	347	281	87	219	292	43	61	0.157	25	170	41	126	203
Kalimantan	411	104	44	153	75	92	38	0.163	19	141	54	118	286
Urban	342	73	50	94	52	95	29	0.177	20	159	58	92	275
Rural	433	114	42	168	84	91	41	0.158	18	135	53	127	290
Sulawesi	361	199	57	150	171	91	41	0.170	20	119	41	129	276
Urban	359	152	51	103	84	100	44	0.210	20	141	40	105	234
Rural	361	205	59	157	184	89	40	0.161	21	114	41	134	285
Maluku & Irian Jaya	231	60	55	258	136	121	30	0.204	22	146	44	137	231
Urban	295	61	41	121	84	108	23	0.237	25	186	54	110	263
Rural	183	60	75	315	173	131	52	0.140	18	118	28	166	208

Source: SUSENAS 1978 data tapes, Biro Pusat Statistik, Jakarta.

Appendix Table

Table A.2: PER CAPITA DAILY CONSUMPTION OF FOODS, FOR HOUSEHOLDS REPORTING CONSUMPTION, BY REGION AND SEASON, INDONESIA, 1978
(in grams except for eggs which are numbers)

Region/Season	Rice	Corn	Wheat	Cassava	Potatoes	Fish	Meat & Poultry	Eggs	Dairy Products	Vegetables	Legumes	Fruits	Other
DKI Jakarta	300	42	19	47	32	29	26	0.236	22	95	62	86	241
February	319	54	18	46	29	33	30	0.236	20	98	68	112	255
May	294	29	21	49	35	28	24	0.247	23	92	56	72	237
November	286	23	19	47	32	26	25	0.224	22	96	62	71	232
West Java	402	86	23	129	69	29	31	0.184	29	136	47	93	234
February	384	81	27	132	76	30	35	0.215	22	143	60	118	265
May	420	174	24	126	69	29	26	0.173	20	137	45	86	228
November	402	60	16	128	63	27	31	0.174	19	127	37	73	207
Central Java	269	267	147	209	130	15	20	0.166	19	151	47	74	202
February	246	242	163	226	113	15	21	0.149	17	180	58	87	236
May	291	323	34	196	139	16	21	0.187	22	150	44	65	191
November	267	267	30	205	134	13	19	0.161	19	122	37	72	178
DI Yogyakarta	209	224	13	207	74	8	14	0.169	15	145	38	53	224
February	201	241	11	155	83	8	13	0.222	11	152	36	54	226
May	205	207	17	197	57	9	12	0.139	10	143	41	50	237
November	221	48	15	245	77	7	17	0.165	20	139	38	55	209
East Java	241	234	101	206	97	22	21	0.148	21	156	50	85	177
February	221	255	116	205	92	24	25	0.146	25	162	55	89	195
May	266	226	95	214	85	24	19	0.147	16	149	51	93	171
November	233	200	77	199	115	19	21	0.149	22	157	44	72	163
Sumatra	406	108	52	179	92	68	36	0.181	18	150	52	112	247
February	404	113	53	184	109	71	37	0.191	19	167	68	142	255
May	413	101	48	184	87	68	35	0.178	17	144	47	102	248
November	401	106	54	168	81	63	35	0.173	17	140	42	92	239
Bali & Nusatenggara	349	277	83	213	281	43	58	0.164	24	169	41	122	202
February	339	301	81	219	362	40	54	0.171	28	170	36	122	219
May	373	310	92	195	279	50	70	0.165	20	189	48	130	204
November	334	187	76	224	230	40	48	0.155	25	145	36	113	183
Kalimantan	411	104	44	153	75	92	38	0.163	19	141	54	118	286
February	414	160	49	142	81	101	38	0.139	18	165	64	152	291
May	406	79	38	154	67	92	43	0.168	19	137	58	100	271
November	412	79	44	165	78	84	34	0.179	19	122	44	97	297
Sulawesi	361	199	57	150	171	91	41	0.170	20	119	41	129	276
February	348	181	67	149	198	89	42	0.166	20	132	36	136	303
May	362	254	53	168	161	93	41	0.176	18	113	44	149	257
November	372	161	52	136	160	90	40	0.169	23	113	42	101	269
Maluku & Irian Jaya	231	60	55	258	136	121	30	0.204	22	146	44	137	231
February	239	63	50	245	118	111	25	0.196	23	151	44	151	247
May	230	53	56	266	158	123	38	0.203	21	147	45	162	233
November	223	66	59	261	133	130	28	0.215	23	140	43	92	213

Source: SUSENAS 1978 data tapes, Biro Pusat Statistik, Jakarta.

Appendix Table

Table A.3: PRICES OF FOODS, BY REGION AND LOCATION, INDONESIA, 1978
(In Rupiah per kilogram, except for eggs)

Region/Location	Rice	Corn	Wheat	Cassava	Potatoes	Fish	Meat & Poultry	Eggs	Dairy Products	Vegetables	Legumes	Fruits	Other
DKI Jakarta	157	104	147	45	127	620	1126	51	864	251	204	206	767
Urban	157	104	147	45	127	620	1126	51	864	251	204	206	767
West Java	140	71	154	24	69	401	989	45	690	88	167	119	369
Urban	152	85	191	32	114	538	1108	49	727	103	181	172	517
Rural	139	70	137	23	60	386	951	43	670	87	165	112	352
Central Java	137	53	92	21	46	291	1004	35	741	66	182	91	779
Urban	153	60	138	28	67	414	1028	40	705	92	183	131	582
Rural	135	53	84	20	42	279	993	33	763	63	181	84	802
DI Yogyakarta	139	42	141	22	56	351	1017	33	757	78	152	114	561
Urban	154	35	142	29	99	519	1002	35	776	84	149	138	893
Rural	137	42	140	21	44	315	1029	32	745	77	152	107	501
East Java	138	66	100	26	52	322	972	34	830	65	172	103	318
Urban	150	65	125	30	90	392	1037	42	811	96	185	140	474
Rural	136	66	97	25	41	313	939	32	853	60	169	93	294
Sumatra	152	69	147	29	106	371	1041	45	647	160	198	115	362
Urban	156	81	147	31	125	382	1157	45	669	168	184	138	444
Rural	152	68	147	28	100	369	1001	45	637	158	202	108	344
Bali & Nusatenggara	133	70	135	36	45	344	602	36	779	96	211	103	339
Urban	140	73	144	46	108	352	660	40	865	95	227	134	578
Rural	132	70	134	36	42	343	594	35	741	96	209	101	320
Kalimantan	148	89	163	32	115	386	983	53	751	165	286	125	428
Urban	153	99	149	38	140	355	1082	56	740	146	243	159	423
Rural	146	86	168	31	106	396	945	51	759	171	309	114	429
Sulawesi	144	62	133	51	76	329	782	41	647	119	232	101	300
Urban	149	70	124	52	94	368	999	47	649	114	260	130	425
Rural	143	60	135	51	73	321	714	39	646	120	225	95	273
Maluku & Irian Jaya	158	102	163	54	140	344	1295	63	678	153	308	152	465
Urban	137	104	167	62	174	426	1415	66	706	156	294	182	579
Rural	173	100	157	51	116	286	872	58	638	152	331	121	383

Source: SUSENAS 1978 data tapes, Biro Pusat Statistik, Jakarta.

Appendix Table

Table A.4: PRICES OF FOODS, BY REGION AND SEASON, INDONESIA, 1978
(In Rupiah per kilogram, except for eggs)

Region/ Season	Rice	Corn	Wheat	Cassava	Potatoes	Fish	Meat & Poultry	Eggs	Dairy Products	Vegetables	Legumes	Fruits	Other
DKI Jakarta	157	104	147	45	127	620	1126	51	864	251	204	206	767
February	154	93	142	49	139	691	1123	52	790	256	199	171	609
May	155	109	137	46	118	618	1076	52	898	261	208	244	912
November	161	127	160	43	127	652	1181	50	898	236	207	206	778
West Java	140	71	154	24	69	401	989	45	690	88	167	119	369
February	148	70	165	26	63	371	899	47	698	93	152	121	346
May	123	78	149	24	69	392	989	45	679	88	162	111	401
November	150	79	138	21	73	439	1047	43	692	83	186	127	362
Central Java	137	53	92	21	46	291	1004	35	741	66	182	91	779
February	147	52	87	26	56	272	939	37	767	60	179	105	1477
May	120	50	140	20	42	290	960	34	699	61	177	92	494
November	147	64	127	17	43	317	1091	33	743	78	189	79	338
DI Yogyakarta	139	42	141	22	56	351	1017	33	757	78	152	114	561
February	148	40	147	28	58	395	949	34	774	63	164	142	462
May	126	44	136	23	71	308	1049	35	809	79	146	103	801
November	145	56	135	17	44	350	1040	31	711	93	147	104	406
East Java	138	66	100	26	52	322	972	34	830	65	172	103	318
February	146	62	101	31	56	326	959	36	737	65	165	105	281
May	121	62	96	25	52	320	962	33	859	71	172	104	315
November	148	77	102	21	47	320	987	33	964	57	177	99	361
Sumatra	152	69	147	29	106	371	1041	45	647	160	198	115	362
February	158	73	145	28	100	340	967	43	648	155	179	110	333
May	139	56	148	28	103	362	1001	44	643	162	205	114	351
November	161	72	150	30	113	412	1134	48	650	163	209	120	402
Bali & Nusatenggara	133	70	135	36	45	344	602	36	779	96	211	103	339
February	142	73	123	42	41	344	587	36	818	133	210	116	336
May	122	66	142	32	56	339	574	34	732	70	209	101	290
November	133	73	139	36	42	348	649	36	797	85	214	94	396
Kalimantan	148	89	163	32	115	386	983	53	751	165	286	125	428
February	146	81	161	32	98	354	975	51	769	148	269	129	384
May	143	92	178	32	120	365	880	52	759	176	284	131	407
November	153	93	155	33	123	437	1071	55	731	170	301	117	491
Sulawesi	144	62	133	51	76	329	782	41	647	119	232	101	300
February	154	67	125	56	82	301	854	41	628	104	237	105	259
May	140	50	131	47	67	330	724	38	689	131	224	93	309
November	139	67	140	50	78	356	782	43	622	122	236	105	330
Maluku & Irian Jaya	158	102	163	54	140	344	1295	63	678	153	308	152	465
February	152	96	167	53	154	358	1340	61	690	148	302	163	444
May	155	100	162	51	125	346	1096	63	659	144	302	145	431
November	166	128	160	59	142	327	1399	67	682	169	322	148	520

Source: SUSENAS 1978 data tapes, Biro Pusat Statistik, Jakarta.

Appendix Table

Table A.5: PER CAPITA DAILY CONSUMPTION OF NUTRIENTS, BY REGION AND LOCATION, INDONESIA, 1978

Region/ Location	Calories (calories)	Protein (grams)	Fat (grams)	Carbohydrates (grams)	Calcium (mg.)	Iron (mg.)	Vitamin A (int. unit)	Thiamine (mg.)	Riboflavin (mg.)	Niacin (mg.)	Vitamin C (mg.)
DKI Jakarta	1854	54	42	320	324	9	4863	0.9	0.7	13	130
Urban	1854	54	42	320	324	9	4863	0.9	0.7	13	130
West Java	2082	54	27	406	264	9	4559	0.9	0.7	16	140
Urban	1971	56	35	360	321	9	5604	0.9	0.7	15	137
Rural	2095	54	26	411	257	9	4433	0.9	0.7	16	140
Central Java	1556	39	23	309	276	9	5750	0.8	0.6	11	146
Urban	1593	45	28	299	300	9	6107	0.9	0.6	11	126
Rural	1552	38	22	310	274	9	5709	0.8	0.7	11	149
DI Yogyakarta	1488	34	25	292	280	9	4122	0.8	0.6	9	120
Urban	1486	40	28	277	255	9	4348	0.9	0.5	10	98
Rural	1488	33	24	294	284	9	4081	0.7	0.6	9	124
East Java	1639	42	27	319	303	10	5037	0.9	0.7	11	120
Urban	1669	46	35	300	297	9	4853	0.9	0.7	11	104
Rural	1634	41	26	322	304	10	5064	0.8	0.7	11	122
Sumatra	2401	63	43	443	346	12	9282	1.0	-	-	246
Urban	2097	59	43	373	347	11	8885	1.0	-	-	251
Rural	2469	63	43	459	345	12	9370	1.1	-	-	245
Bali & Nusatengara	2234	55	35	429	282	11	11946	1.1	0.8	16	153
Urban	2016	53	32	380	251	9	5129	0.9	0.7	15	102
Rural	2252	55	36	433	285	12	12486	1.1	0.8	17	157
Kalimantan	2420	68	37	454	358	11	7426	1.0	0.8	20	166
Urban	2139	65	37	389	411	11	6978	1.0	0.8	17	157
Rural	2512	68	37	476	341	12	7573	1.0	0.8	20	169
Sulawesi	2234	62	37	416	255	10	6071	0.9	-	-	116
Urban	2119	61	34	393	248	10	5793	0.9	-	-	105
Rural	2259	62	38	421	257	10	6134	0.9	-	-	118
Maluku & Irian Jaya	1996	56	43	350	349	10	8697	0.9	0.8	15	215
Urban	2150	64	47	373	405	11	8530	1.0	0.9	17	189
Rural	1886	51	41	333	308	10	8817	0.8	0.7	14	234

Source: SUSENAS 1978 data tapes, Biro Pusat Statistik, Jakarta.

Appendix Table

Table A.6: PER CAPITA DAILY CONSUMPTION OF NUTRIENTS, BY REGION AND SEASON, INDONESIA, 1978

Region/ Season	Calories (calories)	Protein (grams)	Fat (grams)	Carbohydrates (grams)	Calcium (mg.)	Iron (mg.)	Vitamin A (int. unit)	Thiamine (mg.)	Riboflavin (mg.)	Niacin (mg.)	Vitamin C (mg.)
DKI Jakarta	1854	54	42	320	324	9	4863	0.9	0.7	13	130
February	1975	56	42	341	348	10	5342	0.9	0.7	14	135
May	1795	51	42	312	307	8	4298	0.8	0.6	12	117
November	1760	53	42	306	317	9	4966	0.9	0.7	13	139
West Java	2082	54	27	406	264	9	4559	0.9	0.7	16	140
February	2097	56	29	409	284	10	6172	1.0	0.7	17	146
May	2128	55	26	411	262	9	3743	0.9	0.7	16	141
November	2065	52	25	397	244	8	3685	0.8	0.6	16	132
Central Java	1556	39	23	309	276	9	5750	0.8	0.6	11	146
February	1552	41	25	299	314	10	6767	0.9	0.7	11	173
May	1503	39	21	307	260	9	5619	0.8	0.6	11	137
November	1673	36	21	324	254	8	4780	0.7	0.6	10	128
DI Yogyakarta	1488	34	25	292	280	9	4122	0.8	0.6	9	120
February	1442	34	25	271	268	9	5199	0.8	0.6	9	118
May	1405	34	25	279	269	9	3373	0.7	0.5	9	122
November	1621	34	25	327	303	9	3823	0.8	0.6	9	121
East Java	1639	42	27	319	303	10	5037	0.9	0.7	11	120
February	1650	43	30	315	319	10	5668	0.9	0.7	11	118
May	1649	42	26	320	289	10	4644	0.9	0.7	11	122
November	1661	40	25	323	402	9	4783	0.8	0.7	11	118
Sumatra	2401	63	43	443	346	12	9282	1.0	-	-	246
February	2434	66	45	455	382	13	11668	1.1	-	-	262
May	2372	62	43	441	333	11	7780	1.0	-	-	242
November	2332	60	41	434	322	11	8399	1.0	-	-	235
Bali & Nusatenggara	2234	55	35	429	282	11	11946	1.1	0.8	16	153
February	2396	56	40	440	281	12	13713	1.2	0.8	16	157
May	2270	61	37	448	289	12	11096	1.2	0.8	18	165
November	2062	48	28	398	277	10	11004	0.9	0.7	15	137
Kalimantan	2420	68	37	454	358	11	7426	1.0	0.8	20	166
February	2485	71	38	470	390	13	10881	1.1	0.8	21	200
May	2344	66	35	439	344	11	5712	1.0	0.8	19	157
November	2411	65	37	455	340	11	5744	1.0	0.7	19	142
Sulawesi	2234	62	37	416	255	10	6071	0.9	-	-	116
February	2274	62	37	419	266	11	8319	1.0	-	-	115
May	2183	62	37	411	249	10	5575	0.9	-	-	121
November	2193	61	38	418	251	10	4396	0.9	-	-	111
Maluku & Irian Jaya	1996	56	43	350	349	10	8697	0.9	0.8	15	215
February	1983	56	44	351	358	10	9843	0.9	0.8	15	206
May	2052	58	43	362	366	11	10658	0.9	0.8	15	234
November	1911	56	42	337	322	10	5609	0.8	0.8	15	204

Source: SUSENAS 1978 data tapes, Biro Pusat Statistik, Jakarta.

Annex 1:   GROUPING OF FOODS

The grouping of foods is as follows:

(1) Rice includes free-market, self-produced, and glutinous rice, as well as rice byproducts (such as rice flour).

(2) Corn includes both fresh and dried corn on the husk, shelled corn, and corn meal.

(3) Wheat includes wheat flour and other grains.

(4) Cassava refers to fresh and dried cassava, and cassava meal.

(4) Potatoes cover sweet potatoes, potatoes, taro and sago. (The latter two are starchy plants similar to potatoes.)

(6) Fish includes fresh ocean and inland fish, salted and dried fish, canned fish, shrimp, and shellfish.

(7) Meat and poultry covers beef from cattle and carabao (water buffalo), mutton, pork, preserved meat, veal, chicken, and turkey.

(8) Eggs are treated separately because of the difficulty of combining this item with any other.

(9) Dairy products cover fresh milk, evaporated milk, powdered milk, and cheese.

(10) Vegetables include spinach, kangkung spinach, cabbage, mustard greens, beans, peas, strong beans, tomatoes, radishes, carrots, cucumbers, cassava leaves, eggplant, bean sprouts, squash, red and white onions, red and cayenne peppers, and papaya leaves.

(11) Legumes refer to peanuts, green and red beans, soybeans, tunggak beans, bean curd, soybean cake, tauco, peanut cake, and lamtoro.

(12) Fruits include citrus fruits, mangoes, apples, avocados, rambutan, dukuh, durian, salak, pineapple, bananas, papaya, jambu, guava, sawo, belimbing, kedongdong, and water melon.

(13) "Other" includes salt, pepper and other spices, fish paste, ketchup, coconut, cooking oil, butter, lard, brown and granulated sugar, tea, coffee, cocoa, fried fish sticks, noodles, monosodium glutamate, lemon syrup, bottled drinks (lemonade, cola, orange soda, etc.), bread, beer, and alcohol.

For computational purposes, those foods given in units other than kilograms were converted to kilograms. All items in food groups 1-6 and 10-12 were given initially in kilograms; conversion for foods in the remaining groups are given below.

Group 7 contains chicken and turkey, each given in units of one bird. A chicken was assumed to weigh one kilogram; a turkey was assigned a weight of 4.5 kilograms.

Group 8 comprises only eggs; therefore, no conversion was necessary.

In group 9, the item "fresh milk" was given in liters. One liter of fresh milk has a weight of 1.032 kilograms.

Because of the diverse nature of group 13, the foods it contains are presented in several different units. Fried fish sticks, noodles, and bread were the only items given in kilograms. A number of items were given in units of 100 grams which have been converted into kilograms. These items are salt, pepper and other spices, fish paste, butter, lard, brown sugar, granulated sugar, tea, coffee, cocoa, and monosodium glutamate. A bottle of ketchup was assigned a weight of 500 grams; likewise, a coconut was assumed to weigh 500 grams. Cooking oil, lemon syrup, and bottled drinks were each given in liters; the first has a weight of 0.93 kilograms per liter, while the latter two each weighs 1.04 kilograms per liter. A 12-ounce bottle of beer has a weight of 0.36 kilograms; a "shot" of alcohol containing one and a half ounces weighs 0.042 kilograms.

Annex 2: PROBLEMS ASSOCIATED WITH DEFICIENCIES OF NUTRIENTS

Nutrient	Problems associated with deficiency of nutrient
Vitamin A	Poor growth, night blindness, eye damage, xerophthalmia.
Thamine (Vitamin B_1) & Riboflavin (Vitamin B_2)	Beri-beri, weakness of the nervous system, lesions of skin and tongue, eye damage.
Niacin	Pellagra, sometimes with mental disorder related to schizophrenia
Ascorbic acid (Vitamin C)	Scurvy
Iron (associated with Vitamin B_{12})	Anemia

REFERENCES

Chernichovsky, Dov, "The Demand for Nutrition: an Economist's Perspective", Mimeo, Development Economics Department, World Bank, Washington, D.C., 1977.

Lancaster, Kelvin J., "A New Approach to Consumer Theory", Journal of Political Economy, 74, April 1966, 132-157.

Muellbauer, J., "Household Composition Engel Curves and Welfare Comparisons Between Households," European Economic Review, No. 5, June 1974, pp. 103-122.

Reutlinger, Shlomo and Selowsky, Marcelo, Malnutrition and Poverty: Magnitude and Policy Options, Johns Hopkins for the World Bank, Occasional Papers No. 23, Baltimore, 1979.

Thiel, H., "Quantities, Prices and Budget Enquiries," Review of Economic Studies 19, No. 3, 1952, 129-147.

World Bank Publications of Related Interest

Benefits and Costs of Food Distribution Policies: The India Case
Pasquale L. Scandizzo and Gurushri Swamy

Staff Working Paper No. 509. 1982. 54 pages.

ISBN 0-8213-0011-3. Stock No. 0509. $3.

Confronting Urban Malnutrition: The Design of Nutrition Programs
James E. Austin

Describes a framework for systematically carrying out urban nutrition programs that examines several key considerations in nutrition education, on-site feeding, take-home feeding, nutrient-dense foods, ration shops, food coupons, fortification, direct nutrient dosage, and food processing and distribution.

The Johns Hopkins University Press, 1980. 136 pages.

LC 79-3705. ISBN 0-8018-2261-0, Stock No. JH 2261, $6.50 paperback.

Economics of Supplemental Feeding of Malnourished Children: Leakages, Costs, and Benefits
Odin K. Knudsen

Staff Working Paper No. 451. 1981. iv + 76 pages.

Stock No. WP 0451. $3.

Food Distribution and Nutrition Intervention: The Case of Chile
Lloyd Harbert and Pasquale L. Scandizzo

Staff Working Paper No. 512. 1982. 50 pages (including bibliography, annex).

ISBN 0-8213-0001-6. Stock No. WP 0512. $3.

Prices subject to change without notice and may vary by country.

NEW

Food Policy Analysis
C. Peter Timmer, Walter P. Falcon, and Scott R. Pearson

An innovative attempt to link issues of food production and food consumption. Stresses the role of markets and marketing while placing the hunger-problem squarely in a macroeconomic context.

Notes that the solutions to the problem of hunger lie in understanding the food system—the processes that produce agricultural commodities on farms, transform those commodities into foods, and market them to satisfy the nutritional as well as the social and esthetic needs of consumers. Analyzes: the behavior of food consumers and producer households; the effects of macroeconomic forces on the performance of the food system; and the role of markets, both domestic and international, in linking household issues in the micro sector to policy issues in the macro economy.

The Johns Hopkins University Press, 1983. 301 pages.

ISBN 0-8018-3072-9. $25 hardcover; Stock No. JH 3072, $12.95 paperback.

Food Policy Issues in Low-Income Countries
Edward Clay and others

Staff Working Paper No. 473. 1981. 120 pages.

Stock No. WP 0473. $5.

Food Security in Food Deficit Countries

Staff Working Paper No. 393. 1980. 39 pages (including appendix, references).

Stock No. WP 0393. $3.

NEW

International Finance for Food Security
Barbara Huddleston, D. Gale Johnson, Shlomo Reutlinger, and Alberto Valdes

Examines the economic justification for creating an international facility to finance a portion of the food imports to developing countries when special needs arise. A stochastic simulation model is used to analyze the likely stabilizing effect and benefits of the financial facility for different policy scenarios and countries. Evaluates the possible contribution of such a facility and assesses the merits and the probable effects of a facility that has been adopted by the International Monetary Fund.

The Johns Hopkins University Press. 1984. 96 pages.

LC 83-48109. ISBN 0-8018-3070-2. Stock No. JH 3070. $15 hardcover.

Malnourished People: A Policy View
Alan Berg

Discusses the importance of adequate nutrition as an objective, as well as a means of economic development. Outlines the many facets of the nutrition problem and shows how efforts to improve nutrition can help alleviate much of the human and economic waste in the developing world.

1981. 108 pages (including 6 appendixes, notes).

Stock Nos. BK 9029 (English), BK 9030 (French), BK 9031 (Spanish). $5.

Malnutrition and Poverty: Magnitude and Policy Options
Shlomo Reutlinger and Marcelo Selowsky

The first large research effort in the World Bank to determine the global dimension of malnutrition.

The Johns Hopkins University Press, 1976; 2nd printing, 1978. 94 pages (including 5 appendixes).

LC 76-17240. ISBN 0-8018-1868-0, Stock No. JH 1868. $4.75 paperback.

Spanish: Desnutrición y pobreza: magnitudes y opciones de política. Editorial Tecnos, 1977.

ISBN 84-309-0726-2, $4.75.

Measuring Urban Malnutrition and Poverty: A Case Study of Bogota and Cali, Colombia
Rakesh Mohan, M. Wilhelm Wagner, and Jorge Garcia

Staff Working Paper No. 447. 1981. 80

pages (including bibliography, appendixes). Stock No. WP 0447. $3.

Nutrition and Food Needs in Developing Countries
Odin K. Knudsen and Pasquale L. Scandizzo

Staff Working Paper No. 328. 1979. 73 pages (including 4 appendixes).
Stock No. WP 0328. $3.

Nutritional Consequences of Agricultural Projects: Conceptual Relationships and Assessment Approaches
Per Pinstrup-Andersen

Staff Working Paper No. 456. 1981. 93 pages (including bibliography, appendix).
Stock No. WP 0456. $3.

NEW

Poverty, Undernutrition, and Hunger
Michael Lipton

Focuses on the poor and the ultra poor at nutritional risk. Draws on data from two comparable poor regions in Asia and Africa to identify food-related indicators of and distinctions among poverty levels. Implications for nutrition policy, such as balance among poverty projects, are discussed with supporting tables and graphs.

Staff Working Paper No. 597. 1983. 120 pages.
ISBN 0-8213-0204-3. Stock No. WP 0597. $5.

NEW

Prospects for Food Production and Consumption in Developing Countries
Malcolm D. Bale and Ronald C. Duncan

Food production is expected to continue to improve in developing countries, along with consumption of various food items. This study targets pricing policies as critical factors for maximizing agricultural systems in developing countries. Looks at recent world food production and extends trends based on Bank projections to 1995.

Staff Working Paper No. 596. 1983. 40 pages.
Stock No. WP 0596. $3.

NEW

Targeting Food Subsidies for the Needy: The Use of Cost-Benefit Analysis and Institutional Design
Abel Mateus

Analyzes schemes for targeting food subsidies to nutritionally needy groups. Draws lessons from this analysis for setting up and/or reforming current food policy systems. Evaluates ration shops (India, Brazil), self-targeting using an inferior-goods approach (Pakistan, Bangladesh), food-coupon systems (Sri Lanka, Colombia, Indonesia), special intervention programs, and school feeding programs. Provides a critical evaluation of the food subsidy systems that follow a typology reported in this paper.

Staff Working Paper No. 617. 1983. 88 pages.
ISBN 0-8213-0295-7. Stock No. WP 0617. $3.

NEW

Trends in Food and Nutrient Availability in China, 1950-81
Alan Piazza

Project planners in the areas of agriculture and nutrition will find a wealth of practical data in this report. Tables of yearly food balance sheets provide a basis for assessing one nation's ability to feed its people. National and provincial data present generally favorable trends in China.

Staff Working Paper No. 607. 1983. 148 pages.
ISBN 0-8213-0217-5. Stock No. WP 0607. $5.

The World Bank
Publications Order Form

SEND TO: YOUR LOCAL DISTRIBUTOR OR TO   **WORLD BANK PUBLICATIONS**
(See the other side of this form.)   P.O. BOX 37525
   WASHINGTON, D.C. 20013 U.S.A.

Date _____

Name _____   Ship to: (Enter if different from purchaser)

Title _____   Name _____

Firm _____   Title _____

Address_____   Firm _____

City _____ State_____ Postal Code_____   Address_____

Country _____ Telephone (_____) _____   City _____ State_____ Postal Code_____

Purchaser Reference No. _____   Country _____ Telephone (_____) _____

Check your method of payment.
Enclosed is my ☐ Check ☐ International Money Order ☐ Unesco Coupons ☐ International Postal Coupon.
Make payable to World Bank Publications for U.S. dollars unless you are ordering from your local distributor.

Charge my ☐ VISA ☐ MasterCard ☐ American Express ☐ Choice. (Credit cards accepted only for orders addressed
to World Bank Publications.)

_____   _____   _____
Credit Card Account Number   Expiration Date   Signature

☐ Invoice me and please reference my Purchase Order No. _____.

Please ship me the items listed below.

Stock Number	Author/ Title	Customer Internal Routing Code	Quantity	Unit Price	Total Amount $

All prices subject to change.   Prices may vary by country.   Allow 6–8 weeks for delivery.

Subtotal Cost $_____

Total copies _____ Air mail surcharge if desired ($2.00 each) $_____

Postage and handling for more than two complimentary items ($2.00 each) $_____

Total $_____

Thank you for your order.

IBRD-0063

Distributors of World Bank Publications

ARGENTINA
Carlos Hirsch, SRL,
Attn: Ms. Monica Bustos
Florida 165 4° piso
Galeria Guemes
Buenos Aires 1307

AUSTRALIA, PAPUA NEW GUINEA, FIJI, SOLOMON ISLANDS, WESTERN SAMOA, AND VANUATU
The Australian Financial Review
Information Service (AFRIS)
Attn: Mr. David Jamieson
235-243 Jones Street
Broadway
Sydney, NSW 20001

BELGIUM
Publications des Nations Unies
Attn: Mr. Jean de Lannoy
av. du Roi 202
1060 Brussels

CANADA
Le Diffuseur
Attn: Mrs. Suzanne Vermette
C.P. 85, Boucherville J4B 5E6
Quebec

COSTA RICA
Libreria Trejos
Attn: Mr. Hugo Chamberlain
Calle 11-13, Av. Fernandez Guell
San Jose

DENMARK
Sanfundslitteratur
Attn: Mr. Wilfried Roloff
Rosenderns Alle 11
DK-1970 Copenhagen V.

EGYPT, Arab Republic of
Al Ahram
Al Galaa Street
Cairo

FINLAND
Akateeminen Kirjakauppa
Attn: Mr. Kari Litmanen
Keskuskatu 1, SF-00100
Helsinki 10

FRANCE
World Bank Publications
66, avenue d'Iéna
75116 Paris

GERMANY, Federal Republic of
UNO-Verlag
Attn: Mr. Joachim Krause
Simrockstrasse 23
D-5300 Bonn 1

HONG KONG, MACAU
Asia 2000 Ltd.
Attn: Ms. Gretchen Wearing Smith
6 Fl., 146 Prince Edward Road
Kowloon

INDIA
UBS Publishers' Distributors Ltd.
Attn: Mr. D.P. Veer
5 Ansari Road, Post Box 7015
New Delhi 110002
(Branch offices in Bombay, Bangalore, Kanpur, Calcutta, and Madras)

INDONESIA
Pt. Indira Limited
Attn: Mr. Bambang Wahyudi
Jl, Dr. Sam Ratulangi No. 37
Jakarta Pusat

IRELAND
TDC Publishers
Attn: Mr. James Booth
12 North Frederick Street
Dublin 1

ITALY
Licosa Commissionaria Sansoni SPA

Attn: Mr. Giancarlo Bigazzi
Via Lamarmora 45
50121
Florence

JAPAN
Eastern Book Service
Attn: Mr. Terumasa Hirano
37-3, Hongo 3-Chome, Bunkyo-ku 113
Tokyo

KENYA
Africa Book Services (E.A.) Ltd.
Attn: Mr. M.B. Dar
P.O. Box 45245
Nairobi

KOREA, REPUBLIC OF
Pan Korea Book Corporation
Attn: Mr. Yoon-Sun Kim
P.O. Box 101, Kwanghwamun
Seoul

MALAYSIA
University of Malaya Cooperative
Bookshop Ltd.
Attn: Mr. Mohammed Fahim Htj
Yacob
P.O. Box 1127, Jalan Pantai Baru
Kuala Lumpur

MEXICO
INFOTEC
Attn: Mr. Jorge Cepeda
San Lorenzo 153-11, Col. del Valle,
Deleg. Benito Juarez
03100 Mexico, D.F.

MIDDLE EAST
Middle East Marketing Research
Bureau
Attn: Mr. George Vassilou
Mitsis Bldg. 3
Makarios III Avenue
Nicosia
Cyprus
(Branch offices in Bahrain, Greece, Morocco, Kuwait, United Arab Emirates, Jordan)

NETHERLANDS
MBE BV
Attn: Mr. Gerhard van Bussell
Noorderwal 38,
7241 BL Lochem

NORWAY
Johan Grundt Tanum A.S.
Attn: Ms. Randi Mikkelborg
P.O. Box 1177 Sentrum
Oslo 1

PANAMA
Ediciones Libreria Cultural Panamena
Attn: Mr. Luis Fernandez Fraguela R.
Av. 7, Espana 16
Panama Zone 1

PHILIPPINES
National Book Store
Attn: Mrs. Socorro C. Ramos
701 Rizal Avenue
Manila

PORTUGAL
Livraria Portugal
Attn: Mr. Antonio Alves Martins
Rua Do Carmo 70-74
1200
Lisbon

SAUDI ARABIA
Jarir Book Store
Attn: Mr. Akram Al-Agil
P.O. Box 3196
Riyadh

SINGAPORE, TAIWAN, BURMA
Information Publications Private, Ltd.
Attn: Ms. Janet David
02-06 1st Floor, Pei-Fu Industrial
  Building
24 New Industrial Road
Singapore

SPAIN
Mundi-Prensa Libros, S.A.

Attn: Mr. J.M. Hernandez
Castello 37
Madrid

SRI LANKA AND THE MALDIVES
Lake House Bookshop
Attn: Mr. Victor Walatara
41 Wad Ramanayake Mawatha
Colombo 2

SWEDEN
ABCE Fritzes Kungl, Hovbokhandel
Attn: Mr. Eide Segerback
Regeringsgatan 12, Box 16356
S-103 27 Stockholm

SWITZERLAND
Librairie Payot
Attn: Mr. Henri de Perrot
6, rue Grenus
1211 Geneva

TANZANIA
Oxford University Press
Attn: Mr. Anthony Theobold
Maktaba Road, P.O. Box 5299
Dar es Salaam

THAILAND
Central Department Store, Head Office
Attn: Mrs. Ratana
306 Silom Road
Bangkok

Thailand Management Association
Attn: Mrs. Sunan
308 Silom Road
Bangkok

TUNISIA
Société Tunisienne de Diffusion
Attn: Mr. Slaheddine Ben Hamida
5 Avenue de Carthage
Tunis

TURKEY
Haset Kitapevi A.S.
Attn: Mr. Izzet Izerel
469, Istiklal Caddesi
Beyoglu-Istanbul

UNITED KINGDOM AND NORTHERN IRELAND
Microinfo Ltd.
Attn: Mr. Roy Selwyn
Newman Lane, P.O. Box 3
Alton, Hampshire GU34 2PG
England

UNITED STATES
The World Bank Book Store
600 19th Street, N.W.
Washington, D.C. 20433
(Postal address: P.O. Box 37525
Washington, D.C. 20013, U.S.A.)
Baker and Taylor Company
501 South Gladiola Avenue
Momence, Illinois, 60954
380 Edison Way
Reno, Nevada, 89564
50 Kirby Avenue
Somerville, New Jersey, 08876
Commerce, Georgia 30599

Bernan Associates
9730-E George Palmer Highway
Lanham, Maryland, 20761

Blackwell North America, Inc.
1001 Fries Mill Road
Blackwood, New Jersey 08012

Sidney Kramer Books
1722 H Street, N.W.
Washington, D.C. 20006

United Nations Bookshop
United Nations Plaza
New York, N.Y. 10017

VENEZUELA
Libreria del Este
Attn: Mr. Juan Pericas
Avda Francisco de Miranda, no. 52
Edificio Galipan, Aptdo. 60.337
Caracas 1060-A